To get extra value from this book for no additional cost, go to:

http://www.thomson.com/wadsworth.html

thomson.com is the World Wide Web site for Wadsworth/ITP and is your direct source to dozens of on-line resources. *thomson.com* helps you find out about supplements, experiment with demonstration software, search for a job, and send e-mail to many of our authors. You can even preview new publications and exciting new technologies.

thomson.com: *It's where you'll find us in the future.*

Photo Atlas

FOR ANATOMY AND PHYSIOLOGY

David Morton
Frostburg State University

James W. Perry
University of Wisconsin–Fox Valley

WADSWORTH PUBLISHING COMPANY

I(T)P®   An International Thomson Publishing Company

Belmont, CA • Albany, NY • Bonn • Boston • Cincinnati • Detroit • London • Madrid • Melbourne
Mexico City • New York • Paris • Singapore • Tokyo • Toronto • Washington

Biology Editor	JACK CAREY
Assistant Editor	KRISTIN MILOTICH
Editorial Assistant	MICHAEL BURGREEN
Production Editor	CAROL LOMBARDI
Designer	CLOYCE WALL
Print Buyer	KAREN HUNT
Copy Editor	MARY ROYBAL
Cover Designer	GARY HEAD
Cover Photographs	All photos by DAVID MORTON except cadaver shoulder, which is adapted with permission from *Photographic Atlas of the Human Body* by Drs. B. Vidic and F. R. Suarez. Copyright © 1984 Mosby Year Book, Inc.
Composition and Art Preparation	FOG PRESS
Printer	Courier Corporation

COPYRIGHT © 1998
by Wadsworth Publishing Company
A Division of International Thomson Publishing Inc.

I(T)P

The ITP logo is a registered trademark under license.

Printed in the United States of America

FOR MORE INFORMATION, CONTACT:

Wadsworth Publishing Company
10 Davis Drive
Belmont, California 94002
USA

International Thomson Editores
Campos Eliseos 385, Piso 7
Col. Polanco
11560 México D.F. México

International Thomson Publishing Europe
Berkshire House 168-173
High Holborn
London, WC1V 7AA

International Thomson Publishing GmbH
Königswinterer Strasse 418
53227 Bonn
Germany

Thomas Nelson Australia
102 Dodds Street
South Melbourne 3205
Victoria, Australia

International Thomson Publishing Asia
221 Henderson Road
#05-10 Henderson Building
Singapore 0315

Nelson Canada
1120 Birchmount Road
Scarborough, Ontario
Canada M1K 5G4

International Thomson Publishing Japan
Hirakawacho Kyowa Building, 3F
2-2-1 Hirakawacho
Chiyoda-ku, Tokyo 102
Japan

All rights reserved. No part of this work covered by the copyright hereon may be reproduced or used in any form or by any means—graphic, electronic, or mechanical, including photocopying, recording, taping, or information storage and retrieval systems—without the written permission of the copyright holder.

6 7 8 9 10

ISBN 13: 978-0-534-51716-8
ISBN 10: 0-534-51716-1

Dedicated to Beverly,
Dave's Friend and Lifelong Love

Brief Contents

Microscopy 1
Cells 4
Tissues 9
Integumentary System 30
Skeletal System 32
Articulations 50
Skeletal Muscles 53
Nervous System 61
Receptors 65
Endocrine System 70
Digestive System 72
Respiratory System 83
Circulatory System 86
Lymphatic System 94
Urinary System 97
Reproductive System 100
Cat Dissection 110
Fetal Pig Dissection 132
Physiology 138
Magnetic Resonance Imaging 146
Index 147

Detailed Contents

Microscopy   1
　Compound Microscopes   1
　Dissecting Microscopes   2
　Microscope Parts   3

Cells   4
　Ultrastructure   4
　Mitosis   6
　Bacteria   8

Tissues   9
　Epithelia   9
　Soft Connective Tissues   12
　Hard Connective Tissues   15
　Muscle Tissue   20
　Nervous Tissue   23

Integumentary System   30
　Skin   30

Skeletal System   32
　Skull   33
　Skull Bones   38
　Vertebrae   40
　Sacrum/Sternum/Rib   41
　Pectoral Girdle/Bones of Shoulder Joint/Scapula/Clavicle   42
　Bones of Elbow and Wrist Joints/Bones of Hand   43
　Bones of Arm   44
　Female Pelvic Girdle   45
　Male Pelvic Girdle   46
　Os Coxa   47
　Bones of Knee/Femur/Tibia   48
　Fibula/Bones of Foot   49

Articulations   50
　Shoulder and Elbow   50
　Hip/Developing Joint   51
　Knee   52

Skeletal Muscles   53
　Head   53
　Neck   54
　Upper Trunk and Arm   55
　Forearm and Hand   56
　Abdominal Region   57
　Gluteal Region   58
　Thigh   59
　Leg   60

Nervous System   61
　Human Brain Model   61
　Head/Spinal Cord   62
　Human Brain   63
　Sheep Brain   64

Receptors   65
　Eye   65
　Ear   68
　Cochlea/Taste Buds/Olfactory Epithelium   69

Endocrine System   70
　Pituitary Gland   70
　Pineal Gland/Thyroid Gland/Parathyroid Gland/Adrenal Gland   71

Digestive System   72
　Tongue/Live and Papillae   72
　Salivary Glands   73
　Tooth Development   74
　Abdominal Cavity   75
　Esophagus, Stomach, and Duodenum   76
　Esophagus/Cardiac Stomach   77
　Fundic Stomach   78
　Pyloric Stomach   79
　Small Intestine   80
　Large Intestine and Anal Canal   81
　Liver and Pancreas   82

Respiratory System   83
　Human/Fetal Pig   83
　Larynx   84
　Trachea and Lung   85

Circulatory System   86
　Heart   86
　Aorta   90
　Vena Cava/Blood Vessels   91
　Blood Vessels/Lymphatic Vessel   92
　Blood   93

Lymphatic System   94
　Bone Marrow   94
　Thymus and Spleen   95
　Tonsils, Appendix, and Lymph Node   96

Urinary System   97
　Kidney   97
　Ureter, Bladder, and Urethra   99

REPRODUCTIVE SYSTEM 100
Male 100
Female 104

CAT DISSECTION 110
Skeleton 110
Skinned Cat 111
Skeletal Muscles 112
Oral Cavity/Thoracic Cavity 122
Respiratory and Circulatory Systems 123
Heart/Arteries 124
Abdominal Cavity 125
Urinary System 129
Reproductive System 130

FETAL PIG DISSECTION 132
External Anatomy/Oral Cavity 132
Internal Anatomy 133
Circulatory System 134
Urinary and Reproductive Systems 136
Nervous System 137

PHYSIOLOGY 138
Nervous System 138
Doubly Pithing a Frog 139
Skeletal Muscles 140
Circulatory System 143

MAGNETIC RESONANCE IMAGING 146

INDEX 147

Preface

This comprehensive photo atlas can be used with any textbook in any biology class where students must understand the structures of the human body.

We began producing color photo atlases because of their clear advantages of usefulness and accuracy—not only in the laboratory but as review materials to facilitate students' long-term memory. As in all biology courses, the laboratories in anatomy and physiology are visual experiences, full of images of tissue and organ sections, models, skeletons, and other preparations. This atlas will allow students a comprehensive, ongoing review of specimens at any time or place.

The *Photo Atlas for Anatomy and Physiology* will be a useful supplement to any lab manual, whether commercial or in-house: Most commercial manuals still contain only limited color materials, and although instructor-created manuals are tailored to a specific course, they are often poorly illustrated. This book includes a central core of images common to all anatomy and physiology courses and many others that may help instructors expand their course offerings. We welcome your comments about the balance we've achieved and hope you will use the addresses below to let us know how we can help you further.

A book is conceived and born through the collaborative efforts of the authors and many other creative people. Special thanks to Jack Carey, our publisher at Wadsworth, who continues to guide us with his vision; to Carol Lombardi, whose good humor, understanding, perseverance and organizational skills kept all of us on track; to Kristin Milotich for contributing substantial time and effort; and to the following colleagues who provided valuable feedback during the development of this manuscript: Jean Helgeson, Collin County Community College; Dr. Aaron James, Gateway Community College; Michael Kovacs, Broward Community College; Jane Marks, Paradise Valley Community College; Lewis Milner, North Central Technical College; Betsy Ott, Tyler Junior College; David Parker, North Virginia Community College; Dr. David Smith, San Antonio College; Eric Sun, Macon College; and Richard Symmons, California State University–Hayward.

David Morton *James W. Perry*

About the Authors

DAVE MORTON is Chair of the Biology Department at Frostburg State University. After earning a B.S. in Zoology and teaching junior high school, he attended Cornell University, where he received a Ph.D. with a major in Histology and minors in Physiology and Biochemistry. For more than twenty years he has taught numerous introductory, upper-level, and graduate biology courses at the college/university level. Some of his more interesting research publications describe aspects of iron and fluid balance in vampire bats.

David Morton

Department of Biology
Frostburg State University
Frostburg, MD 21532-1099
phone: 301-687-4355
e-mail: d_morton@fre.fsu.umd.edu

JIM PERRY is the Campus Dean at The University of Wisconsin–Fox Valley, where he is also professor of Biological Sciences and teaches General Botany. His academic training is broad, including a B.S. in Zoology, M.S. in Botany and Zoology, and Ph.D. in Botany and Plant Pathology, all from the University of Wisconsin–Madison. Prior to returning to Wisconsin, he was a faculty member at Frostburg State University, serving as the Chair of Biology and teaching introductory biology courses as well as upper-level offerings in fungi, algae, the plant kingdom and electron microscopy.

James W. Perry

Department of Biological Sciences
University of Wisconsin–Fox Valley
1478 Midway Road, P.O. Box 8002
Menasha, WI 54952-8002
phone: 920-832-2610
e-mail: jperry@uwc.edu

Abbreviations Used in Figure Legends

c.s.	cross section	×	magnification (as compared to actual specimen)
l.s.	longitudinal section		
live	photo taken from a living specimen		
prep. slide	photo taken from a prepared slide		
sec.	section		
t.e.m.	transmission electron micrograph		
w.m.	whole mount		

Note on magnifications: We chose to calculate magnifications of the images in this Photo Atlas as compared to that of the actual specimen, rather than indicating which microscope objective lens or magnification was used when the original photo was taken. For example, if the caption reads "0.50×" the image is one-half the size of the specimen. Likewise, an image listed as "500×" would be five hundred the times the actual size of the specimen.

Compound Microscopes   MICROSCOPY

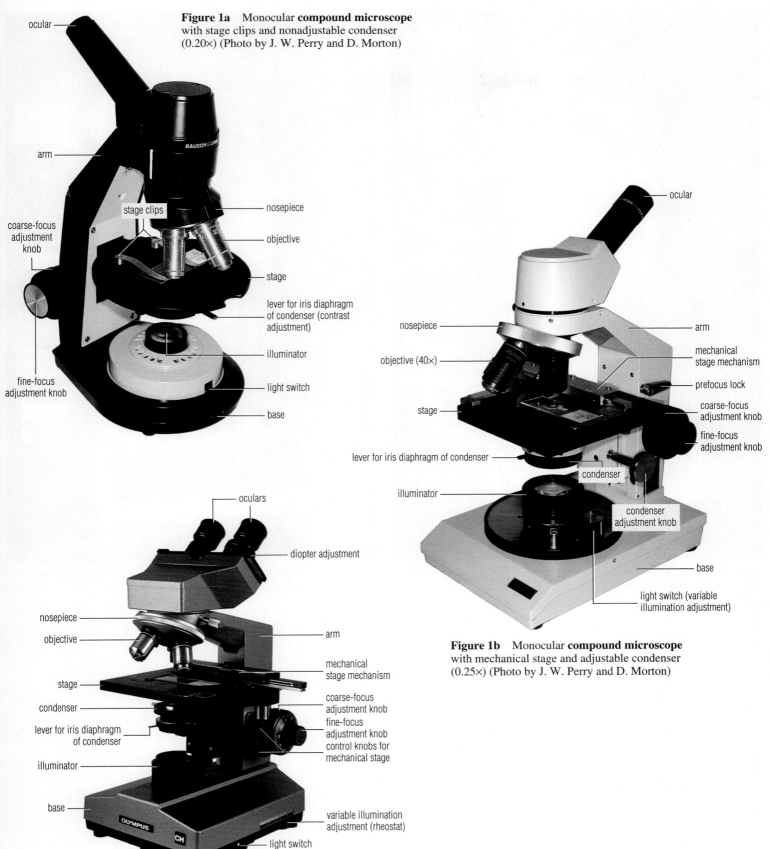

Figure 1a Monocular **compound microscope** with stage clips and nonadjustable condenser (0.20×) (Photo by J. W. Perry and D. Morton)

Figure 1b Monocular **compound microscope** with mechanical stage and adjustable condenser (0.25×) (Photo by J. W. Perry and D. Morton)

Figure 1c Binocular **compound microscope** (0.20×) (Photo by J. W. Perry and D. Morton)

MICROSCOPY *Dissecting Microscopes*

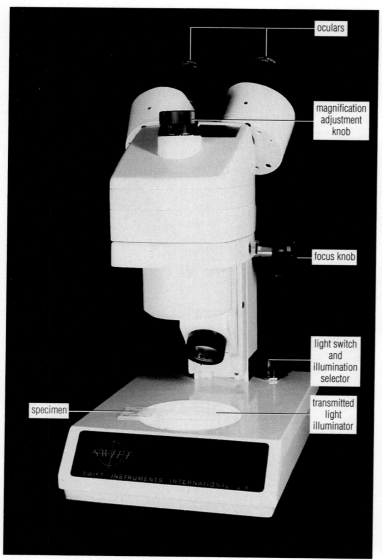

Figure 2a **Dissecting microscope** set up for viewing a transparent slide or other specimen using transmitted light (0.25×). (Photo by D. Morton)

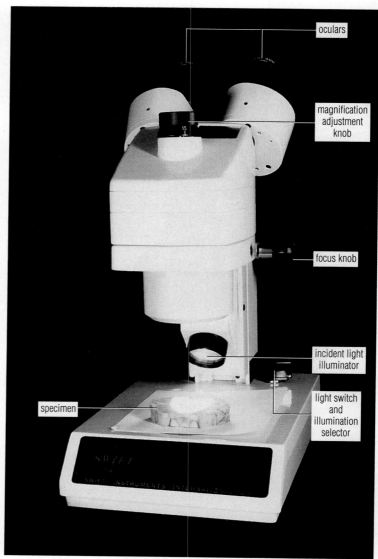

Figure 2b **Dissecting microscope** set up for viewing a specimen using incident or reflected light (0.25×). (Photo by D. Morton)

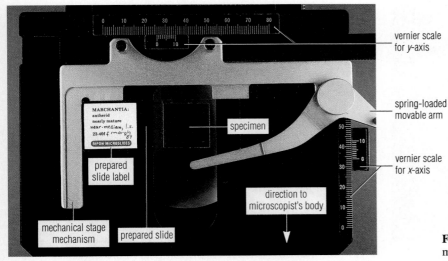

Figure 3a Correct placement of prepared microscope slide on **mechanical stage**. The slide label is oriented so that it can be read by the microscopist (0.50×). (Photo by J. W. Perry)

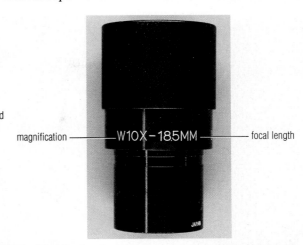

Figure 3d **Ocular** (eyepiece) removed from microscope. The markings on the ocular indicate this to be a wide-field lens with a magnification of 10 (10×) and a focal length of 18.5 mm (1×). (Photo by J. W. Perry)

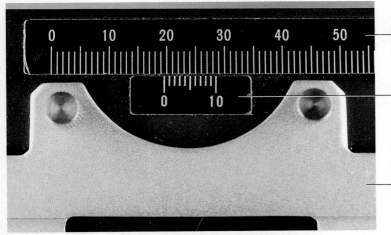

Figure 3b **Vernier scale** on mechanical stage of compound microscope. The correct reading is 19.6 mm (0.60×). (Photo by J. W. Perry)

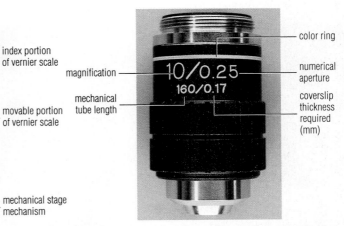

Figure 3e **10× objective**. Engravings indicate that the numerical aperture is 0.25, the mechanical tube length is 160 mm, and a coverslip 0.17 mm thick (No. 2 coverslip) must be used over the specimen for optimal resolution (1×). (Photo by J. W. Perry)

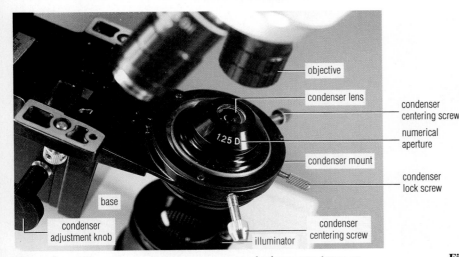

Figure 3c Adjustable **condenser** on a compound microscope (stage removed). The engraved number is the condenser lens's numerical aperture. The letter "D" following the numerical aperture indicates that this condenser lens is to be used dry—that is, with no oil between the lens and the bottom of the slide (0.60×). (Photo by J. W. Perry)

Figure 3f **100× oil immersion objective**. Engravings indicate that the numerical aperture is 1.25, the mechanical tube length is 160 mm, and the objective is to be used with a slide having a coverslip 0.17 mm thick (No. 2 coverslip). The black ring indicates that the gap between the lens and the microscope slide's coverslip is to be filled with immersion oil (1.3×). (Photo by J. W. Perry)

4 CELLS *Ultrastructure*

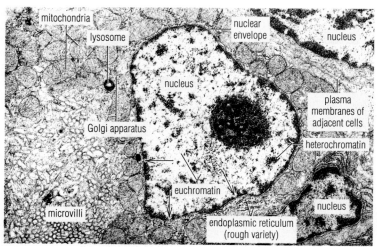

Figure 4a **Ultrastructure** of a parietal cell from the stomach of a common vampire bat with many **cell organelles** typical of **animal cells**. The unlabeled arrows indicate nuclear pores in the nuclear envelope. (t.e.m., 9,000×). (Photo by D. Morton)

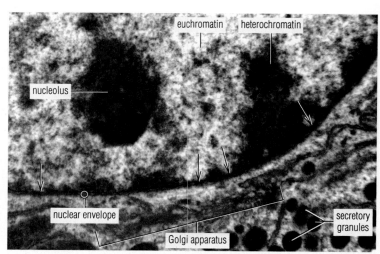

Figure 4b **Nuclear envelope** and **nuclear pores** (arrows) (t.e.m., 25,500×). (Photo by D. Morton)

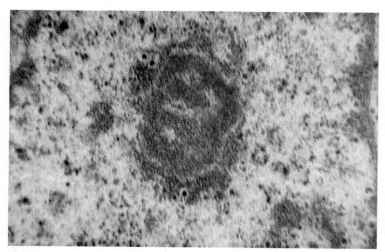

Figure 4c **Nucleolus** (t.e.m., 32,000×). (Photo by D. Morton)

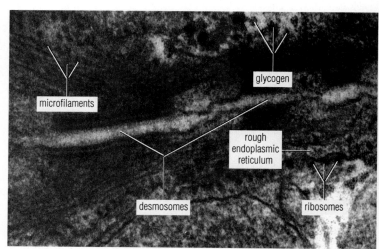

Figure 4d **Rough endoplasmic reticulum** and structural **microfilaments** (t.e.m., 92,500×). (Photo by D. Morton)

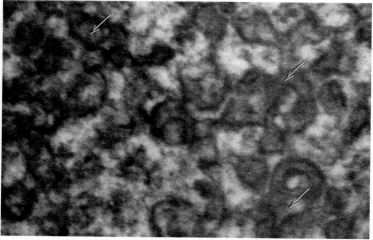

Figure 4e **Smooth endoplasmic reticulum.** The arrows indicate the cisterna (space) between its membranes (t.e.m., 100,000×). (Photo by D. Morton)

Ultrastructure CELLS 5

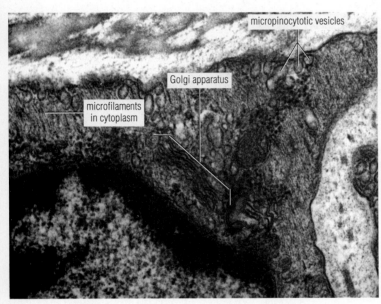

Figure 5a **Golgi apparatus** and contractile **microfilaments** in **smooth muscle cell** (t.e.m., 33,000×). (Photo by D. Morton)

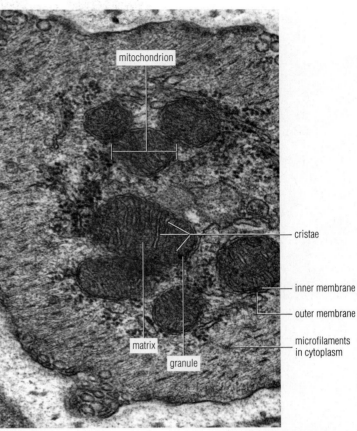

Figure 5b **Mitochondria** in the perinuclear area of a **smooth muscle cell**. The space between the inner and outer membranes is continuous with space within cristae (t.e.m., 33,000×). (Photo by D. Morton)

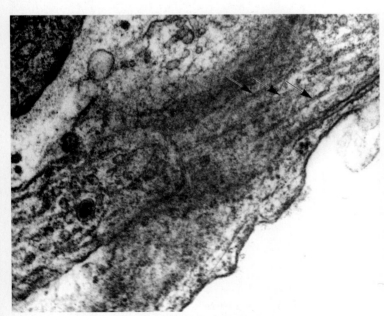

Figure 5c **Microtubules** (arrows) in a process of a fibroblast (t.e.m., 46,000×). (Photo by D. Morton)

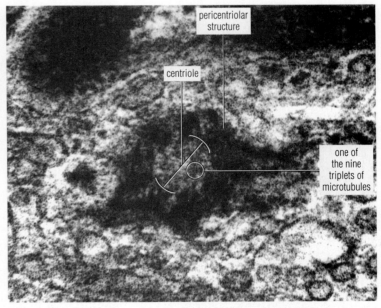

Figure 5d **Centriole** (t.e.m., c.s., 49,500×). (Photo by D. Morton)

6 CELLS *Mitosis*

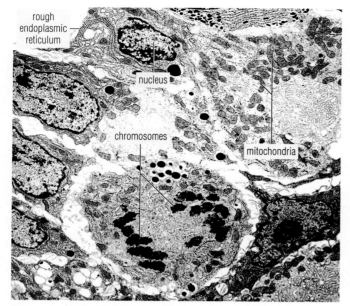

Figure 6a Mitosis (t.e.m., 4100×). (Photo by D. Morton)

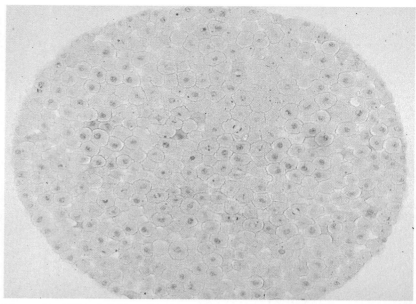

Figure 6b Whitefish blastula. The blastula is a hollow ball of actively dividing cells produced early during embryonic development. Cells illustrated in Figures 5c and 5d are found in this blastula (prep. slide, sec., 100×). (Photo by J. W. Perry)

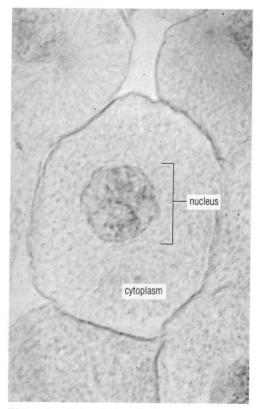

Figure 6c Interphase cell prior to mitosis (prep. slide, sec., 1200×). (Photo by J. W. Perry)

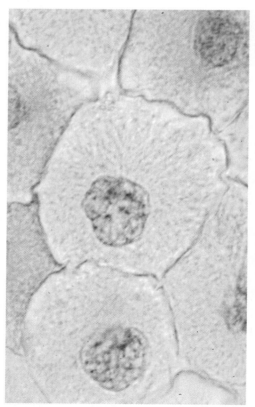

Figure 6d Prophase (prep. slide, sec., 1200×). (Photo by J. W. Perry)

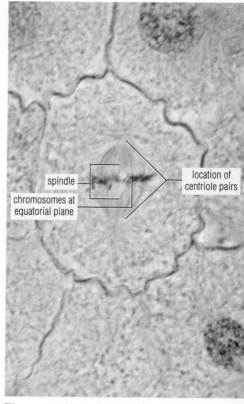

Figure 6e Metaphase (prep. slide, sec., 1200×). (Photo by J. W. Perry)

Mitosis CELLS

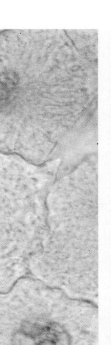

Figure 7a Early anaphase (prep. slide, sec., 1200×). (Photo by J. W. Perry)

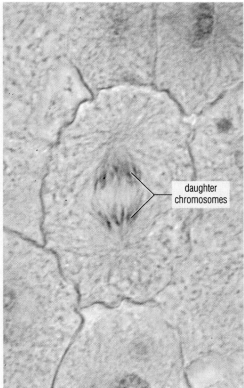

Figure 7b Later anaphase (prep. slide, sec., 1200×). (Photo by J. W. Perry)

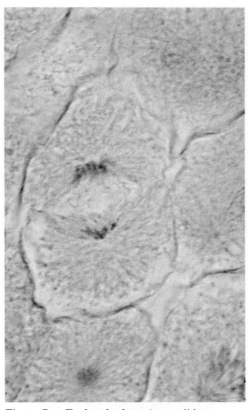

Figure 7c Early telophase (prep. slide, sec., 1200×). (Photo by J. W. Perry)

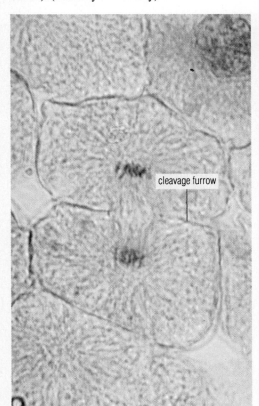

Figure 7d Cytokinesis by **furrowing**, a feature characteristic of most animal cells (prep. slide, sec., 1200×). (Photo by J. W. Perry)

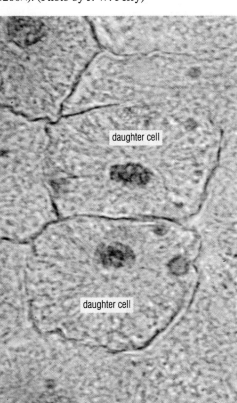

Figure 7e Two **daughter cells** following cytokinesis (prep. slide, sec., 1200×). (Photo by J. W. Perry)

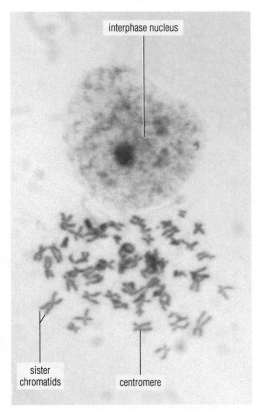

Figure 7f Chromosome spread of **HeLa cells**. Sister chromatids connected at their centromeres are clearly visible (prep. slide, w.m., 90×). (Photo by J. W. Perry)

8 CELLS *Bacteria*

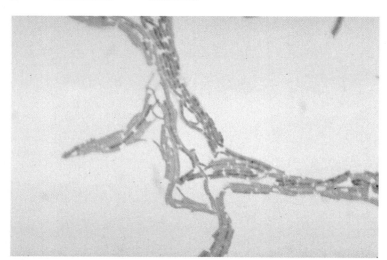

Figure 8a Rodlike **bacilli** (singular, bacillus), one of the three basic shapes of bacteria (prep. slide, w.m., 1000×). (Photo by D. Morton)

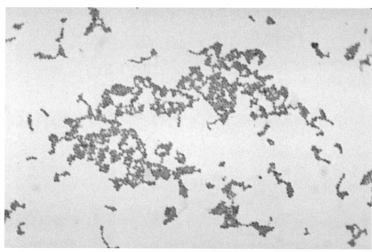

Figure 8b Round bacteria or **cocci** (singular, coccus) (prep. slide, w.m., 1000×). (Photo by D. Morton)

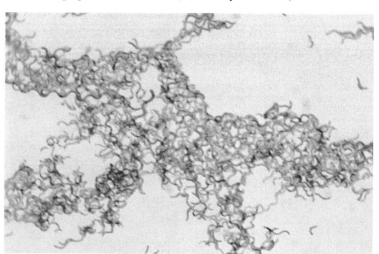

Figure 8c Spiral-shaped **spirilla** (singular, spirillum) (prep. slide, w.m., 1000×). (Photo by D. Morton)

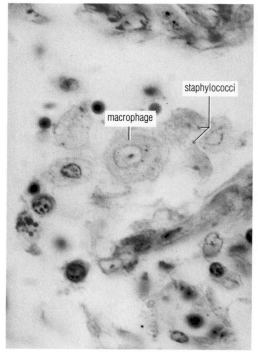

Figure 8d Infection site associated with a lung abscess. **Macrophages** are mononuclear phagocytes, cells that engulf pathogens, other foreign matter, and cellular debris (prep. slide, sec., 1000×). (Photo by D. Morton)

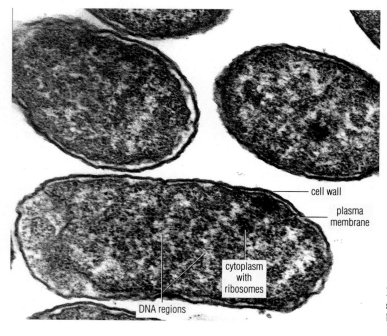

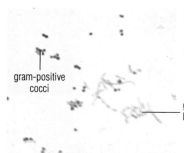

Figure 8f Typical results of a **Gram's stain** on a mixed bacterial culture of *Escherichia coli* and *Staphylococcus aureus* (w.m., 1,200×). (Photo by D. Morton)

Figure 8e *Escherichia coli* **ultrastructure** (t.e.m., 25,000×). (Photo by D. Morton)

Epithelia TISSUES

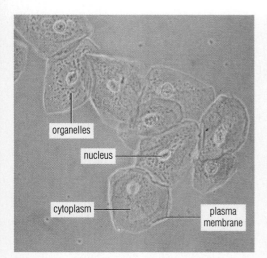

Figure 9a **Unstained human cheek cells** (live, w.m., 500×). (Photo by D. Morton)

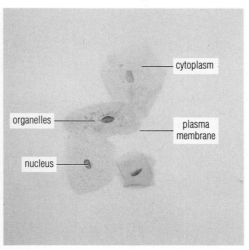

Figure 9b **Human cheek cells stained** with methylene blue (live, w.m., 300×). (Photo by D. Morton)

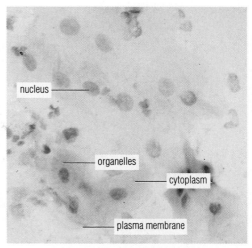

Figure 9c **Human cheek cells** (prep. slide, w.m., 450×). (Photo by D. Morton)

Figure 9d **Simple squamous epithelium** lining one of the surfaces of a piece of **mesentery** that has been placed in a solution of silver nitrate and placed in sunlight. Silver stains the glycoproteins present between adjacent cells. Regardless of their shape when viewed from the side, all epithelial cells are polygonal in surface view (w.m., 300×). (Photo by D. Morton)

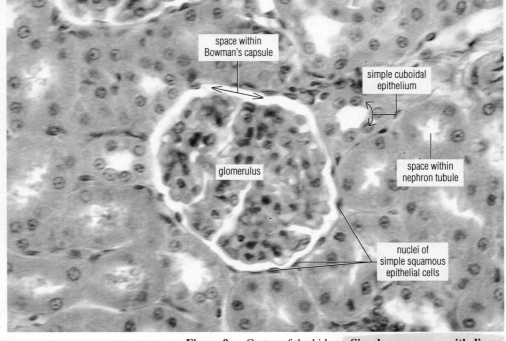

Figure 9e Cortex of the kidney. **Simple squamous epithelium** comprises the outer (parietal) wall of **Bowman's capsule**, which, together with a tight network of capillaries called the **glomerulus**, constitutes a **renal corpuscle**. Squamous cells seen from the side are flattened or scale-like in appearance. Bowman's capsule is the initial portion of the **nephron tubule**. The walls of the other portions of the nephron tubule located in the cortex here seen in cross section are **simple cuboidal epithelium.** Cuboidal cells are cube-like in side view (prep. slide, sec., 350×). (Photo by D. Morton)

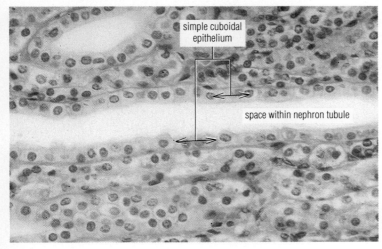

Figure 9f **Simple cuboidal epithelium** (longitudinal section of a nephron tubule) in kidney cortex (prep. slide, sec., 350×). (Photo by D. Morton)

10 TISSUES *Epithelia*

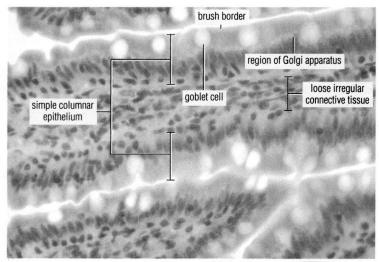

Figure 10a Villi of **small intestine**. **Simple columnar epithelium** lines the surface of the longitudinally sectioned villus in the center of this photo. Columnar cells are shaped like columns in side view. The core of the villus consists largely of **loose irregular connective tissue** (prep. slide, sec., 400×). (Photo by D. Morton)

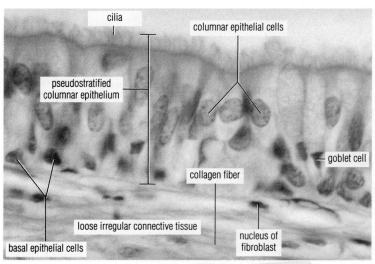

Figure 10b Ciliated **pseudostratified columnar epithelium** lining the lumen of the **trachea**. The several layers of nuclei in this simple epithelium make it appear stratified—hence its name, which means falsely stratified (prep. slide, sec., 1000×). (Photo by D. Morton)

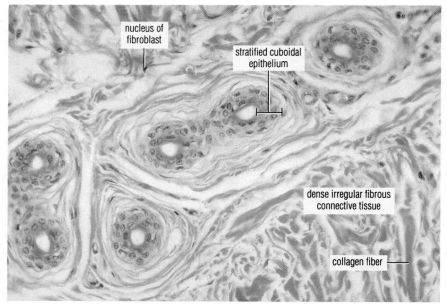

Figure 10c Cross sections of **sweat gland ducts** surrounded by dermis of the skin. The walls of the sweat glands are composed of **stratified cuboidal epithelium**. Both this epithelial subtype and stratified columnar epithelium are relatively rare (prep. slide, sec., 300×). (Photo by D. Morton)

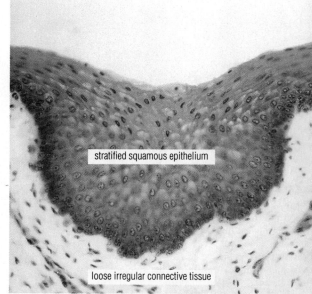

Figure 10d **Stratified squamous epithelium** lining the lumen of the **esophagus**. Stratified epithelia are named according to the shape of the surface cells (prep. slide, sec., 300×). (Photo by D. Morton)

Epithelia TISSUES

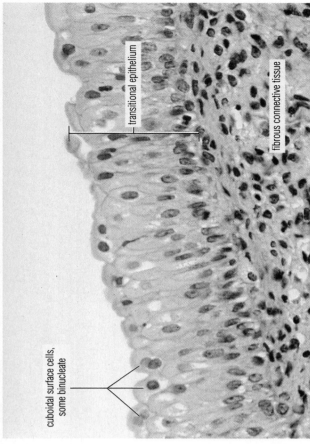

Figure 11b **Transitional epithelium** lining the lumen of a **contracted urinary bladder.** The epithelium will change its appearance depending on the state of stretch of the organ (prep. slide, sec., 950×). (Photo by D. Morton)

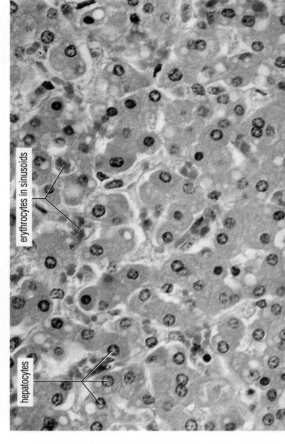

Figure 11d **Parenchyma** or massed cells and ducts of the **liver.** During development, many glands form from an inpocketing of an epithelial surface. In exocrine glands, the ducts connecting the gland to the surface persist in the adult organ. Although difficult to see in a photomicrograph, in the liver there are tiny bile canaliculi between the hepatocytes. These canaliculi drain bile into bile ductules (small ducts), which in turn drain into larger ducts leading to the gall bladder and the lumen of the small intestine (prep. slide, sec., 300×). (Photo by D. Morton)

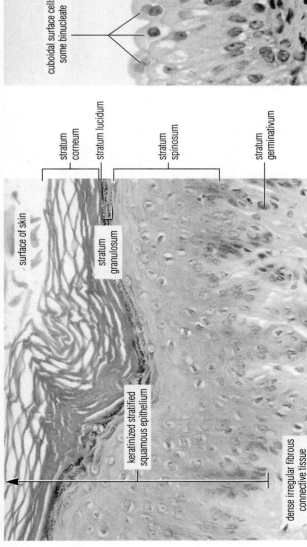

Figure 11a **Keratinized stratified squamous epithelium** in the **epidermis of the skin.** The formation of strata is due to the synthesis of keratin in keratinocytes, which are formed by cell division in the basal stratum and are pushed by the formation of younger cells in two to three weeks to the surface of the upper stratum, from which they are shed (prep. slide, sec., 450×). (Photo by D. Morton)

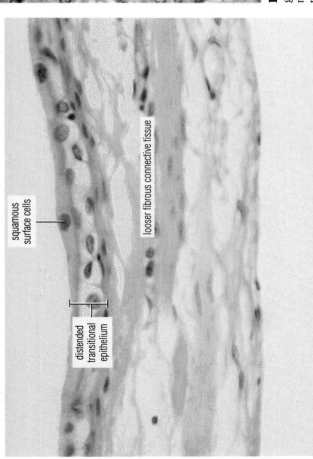

Figure 11c **Transitional epithelium** lining the lumen of a **stretched urinary bladder** (prep. slide, sec., 350×). (Photo by D. Morton)

12 TISSUES Soft Connective Tissues

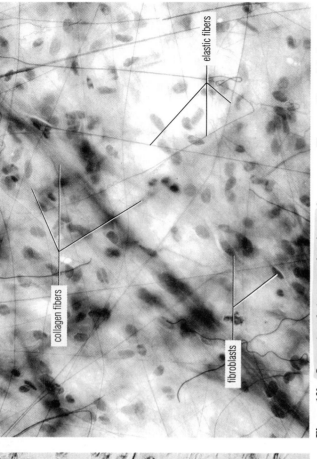

Figure 12a Embryonic connective tissue. The first step in the development of soft connective tissue occurs when mesenchymal cells differentiate into fibroblasts and start secreting collagen molecules. This tissue can be found in embryos and in the umbilical cord and is sometimes called mucous connective tissue (prep. slide, sec., 500×). (Photo by D. Morton)

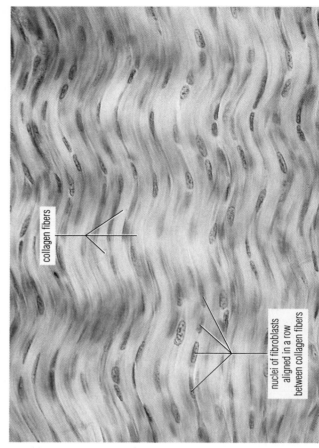

Figure 12b Loose (areolar) connective tissue of mesentery. The term "loose" indicates that fibers are relatively scarce (prep. slide, w.m., 450×). (Photo by D. Morton)

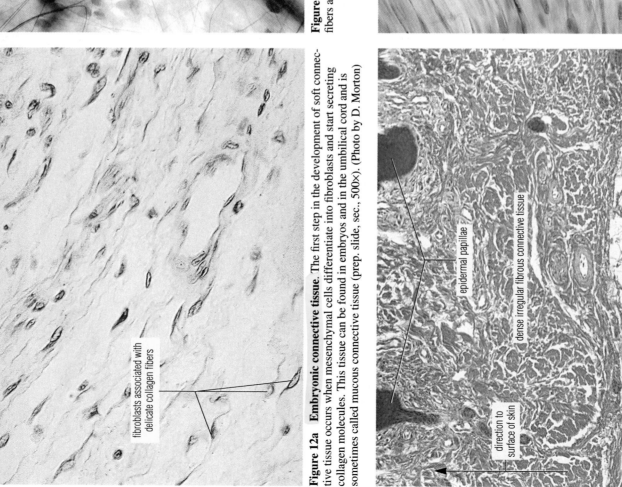

Figure 12c Dense irregular fibrous connective tissue in the **dermis** of the skin. In dense connective tissues, fibers predominate. The term "irregular" refers to the apparently random orientation of the fibers. The term "fibrous" means that most of the fibers are collagen (prep. slide, sec., 300×). (Photo by D. Morton)

Figure 12d Dense regular fibrous connective tissue from a **tendon**. Tendons are structures that connect skeletal muscle organs to bones. Note that all of the collagen fibers are parallel, or regular (prep. slide, sec., 500×). (Photo by D. Morton)

Soft Connective Tissues

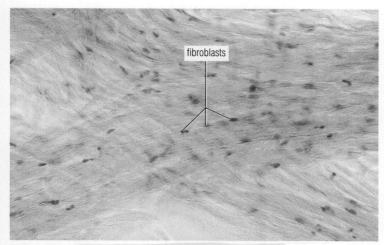

Figure 13a **Dense regular elastic connective tissue** from a **ligament**, which has been teased (pulled apart). Ligaments are structures that connect bones to bones across joints (prep. slide, sec., 450×). (Photo by D. Morton)

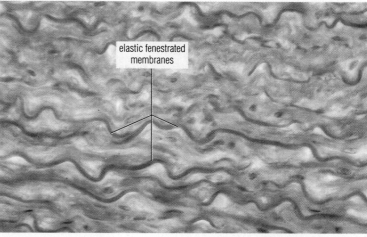

Figure 13b **Elastic fenestrated membranes**. The materials that compose elastic fibers can be assembled into other shapes, such as these holey sheets in the walls of elastic arteries (prep. slide, sec., 450×). (Photo by D. Morton)

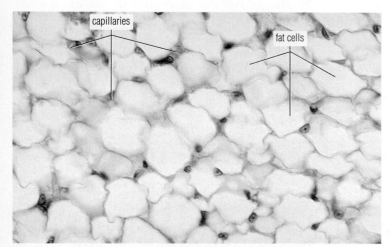

Figure 13c **White adipose tissue**. Not many of the fat cell (or adipocyte) nuclei can be seen because they are "out of the plane of section," which means they are in a piece of these large cells located in a previous or subsequent section. The adipocytes are empty because fat is extracted during standard slide preparation (prep. slide, sec., 450×). (Photo by D. Morton)

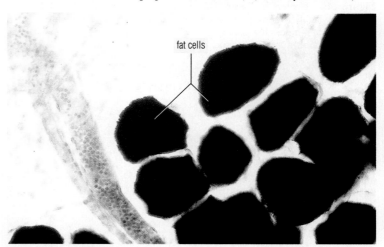

Figure 13d **White adipose tissue** fixed in osmium tetroxide, a fixative that maintains fat and stains it black (prep. slide, sec., 450×). (Photo by D. Morton)

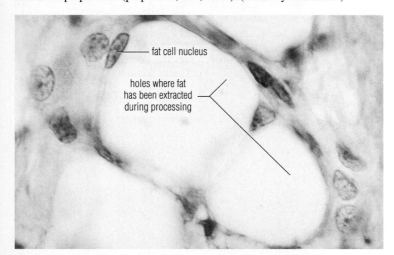

Figure 13e **Unilocular fat cells** with one large fat droplet. Their main function is energy storage and release (prep. slide, 900×). (Photo by D. Morton)

14 TISSUES Soft Connective Tissues

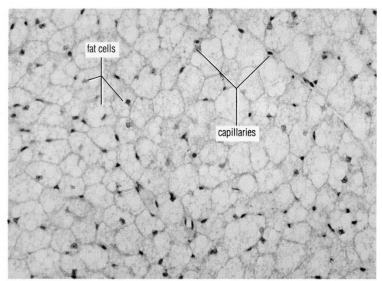

Figure 14a **Brown fat tissue**. This tissue contains multilocular fat cells with many smaller fat droplets. It is common in newborns and in cold-acclimated and hibernating mammals, where its function is thermogenesis (heat production) (prep. slide, sec., 250×). (Photo by D. Morton)

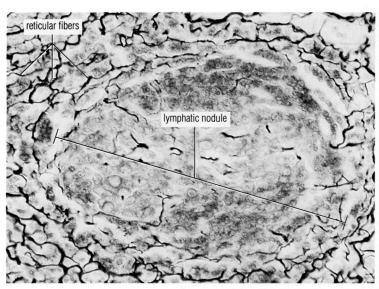

Figure 14b Silver-stained **reticular connective tissue** forming a network of delicate collagen fibers in a **lymph node** (prep. slide, sec., 450×). (Photo by D. Morton)

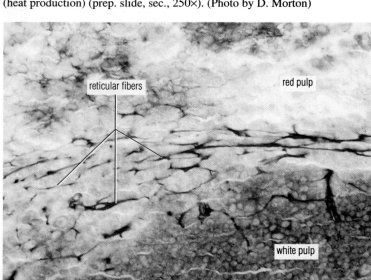

Figure 14c Silver-stained **reticular connective tissue** in the **spleen** (prep. slide, sec., 450×). (Photo by D. Morton)

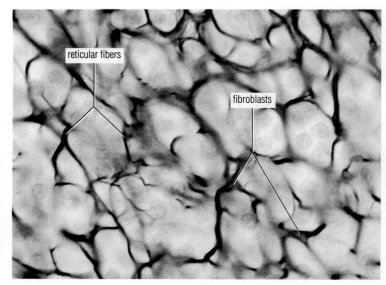

Figure 14d High-power view of **reticular connective tissue**. Although made of the same basic molecules, reticular fibers are much more sensitive to silver staining than collagen fibers (prep. slide, sec., 1100×). (Photo by D. Morton)

Hard Connective Tissues/Cartilage TISSUES

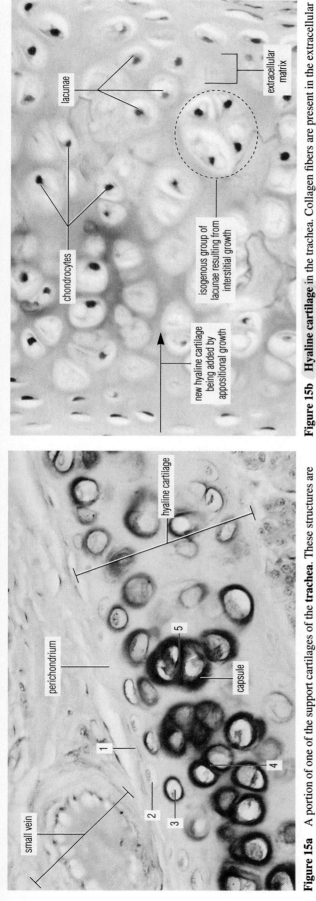

Figure 15a A portion of one of the support cartilages of the **trachea**. These structures are composed of **hyaline cartilage**, a hard connective tissue. The support cartilage is covered by the perichondrium, a soft connective tissue membrane. New hyaline cartilage is added by two growth mechanisms. **Appositional growth** is characterized by chondroblasts (1), present on the inner surface of the perichondrium, secreting extracellular matrix, which forms a lacuna or cavity around the cell (2). Once trapped in the matrix, chondroblasts are called chondrocytes. **Interstitial growth** commences as the chondrocytes continue to secrete matrix (3) and undergo cell division (4), resulting in a related group of cells, or **isogenous group** (5), and the surrounding **capsule** of new matrix fibers (prep. slide, sec., 500×). (Photo by D. Morton)

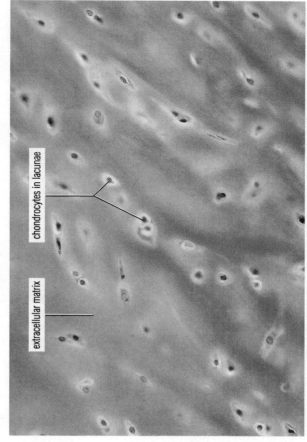

Figure 15b **Hyaline cartilage** in the trachea. Collagen fibers are present in the extracellular matrix, but they cannot be seen because they have the same optical properties as the surrounding polymerized ground substance (prep. slide, sec., 250×). (Photo by D. Morton)

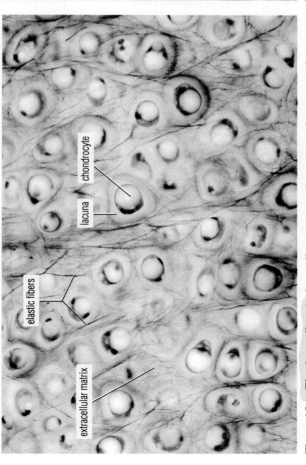

Figure 15c **Elastic cartilage** in the cartilage structure that supports the outer ear flap, or pinna (prep. slide, sec., 300×). (Photo by D. Morton)

Figure 15d **Fibrocartilage** of an intervertebral disc. A higher concentration of collagen fibers is visible in extracellular matrix (prep. slide, sec., 300×). (Photo by D. Morton)

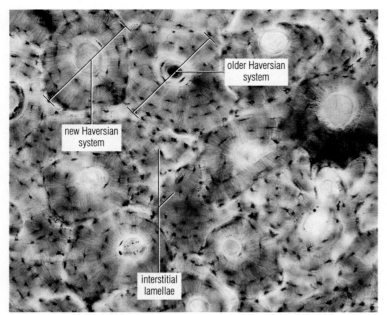

Figure 16a **Bone tissue** in a ground section of dried **compact bone** from the diaphysis (shaft) of a long bone. Compact bone is composed primarily of long cylindrical structures called **Haversian systems** or **osteons**. Periodically, bone tissue is removed and new Haversian systems form in the resulting resorption cavities. Portions of older systems without their central Haversian canal remain as angular areas of **interstitial lamellae** (prep. slide, c.s., 100×). (Photo by D. Morton)

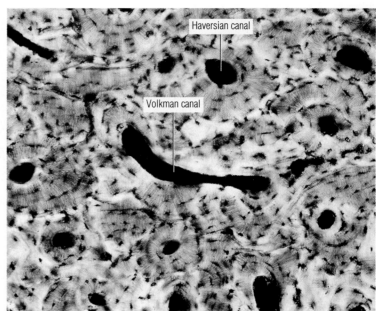

Figure 16b **Haversian canals** are connected to each other by perpendicular **canals of Volkman**. Both types of canals contain capillaries and nerves in living compact bone (prep. slide, c.s., 100×). (Photo by D. Morton)

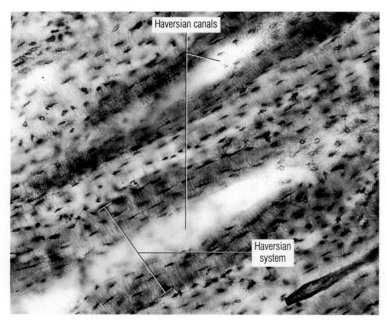

Figure 16c **Haversian canals** seen in **longitudinal view** (prep. slide, l.s., 300×). (Photo by D. Morton)

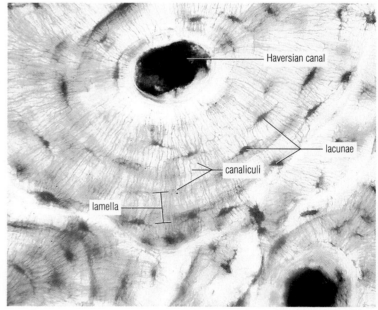

Figure 16d In living compact bone, cells or **osteocytes** reside in rings of **lacunae** oriented around the Haversian canals and separated by **lamellae** of extracellular matrix. Adjacent lacunae are connected to each other and to the canals by **canaliculi** (prep. slide, c.s., 450×). (Photo by D. Morton)

Hard Connective Tissues/Bone/Decalcified

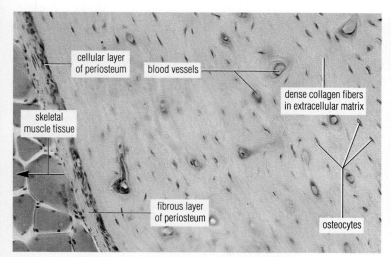

Figure 17a **Decalcified bone tissue** in the outer portion of compact bone from the diaphysis of a long bone. A membranous **periosteum** of dense irregular fibrous connective tissue covers the outer surface of bone organs. The mineral portion of the matrix is absent in decalcified bone, while the cells and collagen fibers remain (prep. slide, c.s., 100×). (Photo by D. Morton)

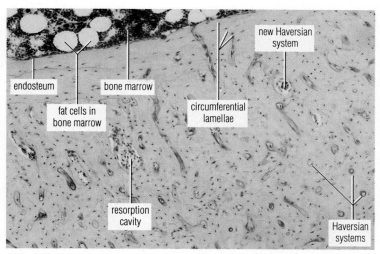

Figure 17b Inner portion of compact bone from the diaphysis of a long bone. A thin membranous **endosteum** covers the surface of the medullary (or bone marrow) cavities within bone organs. Sometimes, **circumferential lamellae** form under the membranes (prep. slide, c.s., 100×). (Photo by D. Morton)

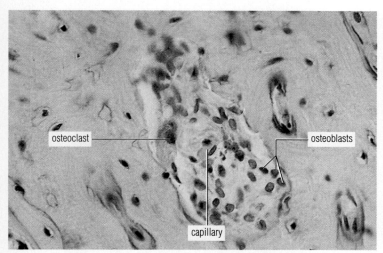

Figure 17c **Resorption cavity** produced by the removal of bone tissue by multinuclear **osteoclasts**. **Osteoblasts** secrete new matrix and once they become trapped within it are called osteocytes (prep. slide, c.s., 250×). (Photo by D. Morton)

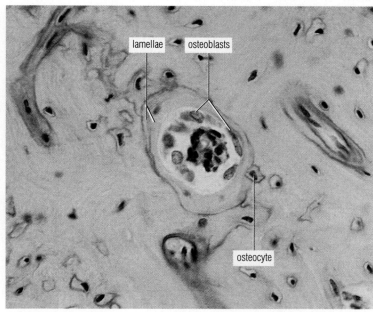

Figure 17d Formation of a **new Haversian system** in a resorption cavity (prep. slide, c.s., 300×). (Photo by D. Morton)

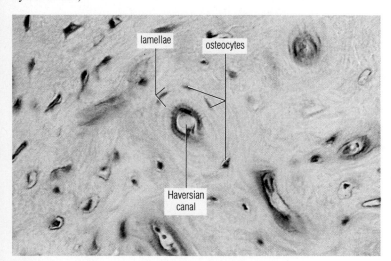

Figure 17e Older **Haversian system** (prep. slide, c.s., 300×). (Photo by D. Morton)

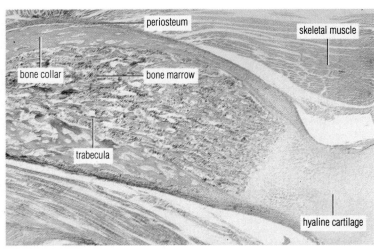

Figure 18a **Primary center of ossification** and **periosteal bone collar** stage of **endochondral ossification**. The center in the middle of the diaphysis of a long bone contains **trabeculae** of **endochondral bone** and **bone marrow** in cavities that will develop into the medullary cavity. Bone organs that ossify by an endochondral mechanism are initially composed of hyaline cartilage (prep. slide, l.s., 30×). (Photo by D. Morton)

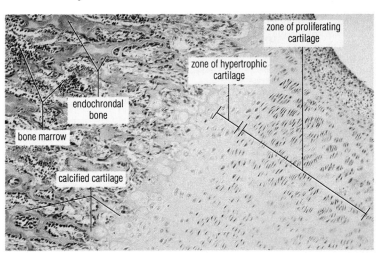

Figure 18b As **endochondral ossification** proceeds, the primary center lengthens as chondrocytes in the remaining hyaline cartilage proliferate and undergo hypertrophy. Then the matrix mineralizes, and the calcified cartilage is removed and replaced by endochondral bone (prep. slide, l.s., 100×). (Photo by D. Morton)

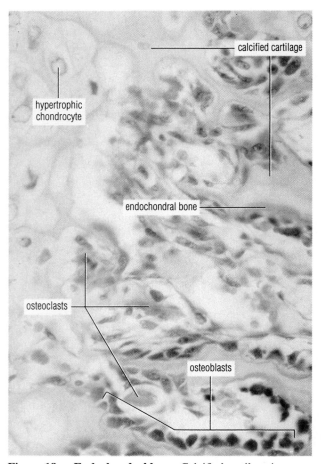

Figure 18c **Endochondral bone**. Calcified cartilage is removed by osteoclasts. This process creates tunnels where osteoblasts secrete bone matrix along the walls (prep. slide, l.s., 300×). (Photo by D. Morton)

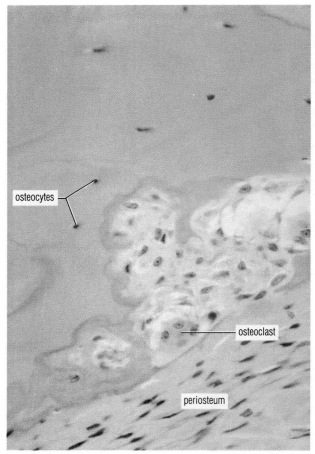

Figure 18d During bone remodeling, **osteoclasts** continue to remove bone tissue (prep. slide, l.s., 300×). (Photo by D. Morton)

Hard Connective Tissues/Bone/Ossification/Endochondral and Intramembranous

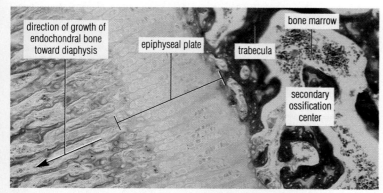

Figure 19a Later in the **endochondral ossification** of a long bone, **secondary centers of ossification** form in the epiphyses (knobby ends), leaving epiphyseal plates of hyaline cartilage, which continues to grow toward the diaphysis and to be replaced by endochondral bone. This process is responsible for the growth in length of bones. At the end of adolescence, the growth plates are closed by the complete replacement of the hyaline cartilage by bone tissue. This slide is stained with a trichrome procedure (prep. slide, l.s., 100×). (Photo by D. Morton)

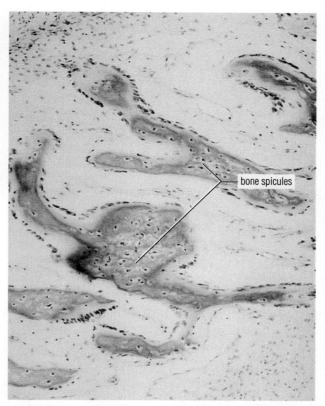

Figure 19b In bones that form by **intramembranous ossification** (e.g., the flat bones of the skull), bone tissue is formed directly in soft connective tissue (prep. slide, sec., 100×). (Photo by D. Morton)

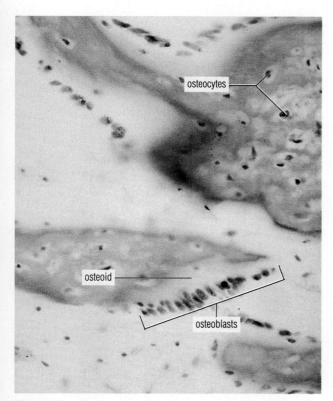

Figure 19c Rows of osteoblasts secrete **osteoid**—the organic components of bone matrix—which is quickly mineralized. Once the osteoblasts are trapped in lacunae in the matrix, they are called osteocytes (prep. slide, sec., 300×). (Photo by D. Morton)

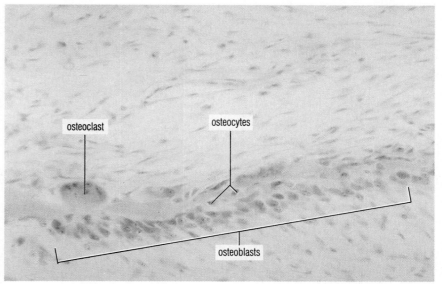

Figure 19d Later in **intramembranous ossification** of the flat bones of the skull, the bone spicules fuse. Osteoblasts lining the outer surface of the bones continue to secrete bone matrix while osteoclasts on the inner side remove bone, enlarging the cranial cavity (prep. slide, sec., 250×). (Photo by D. Morton)

20 TISSUES Muscle Tissue/Skeletal

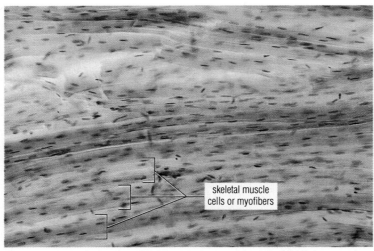

Figure 20a **Skeletal muscle organ** teased to show individual muscle cells with transverse striations and many peripherally located nuclei (prep. slide, sec., 250×). (Photo by D. Morton)

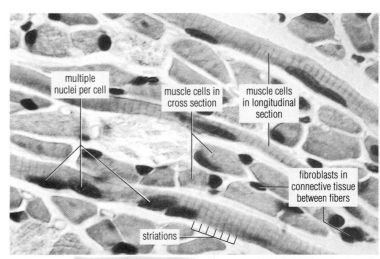

Figure 20b **Skeletal muscle tissue** in a section of the tongue (prep. slide, 400×). (Photo by D. Morton)

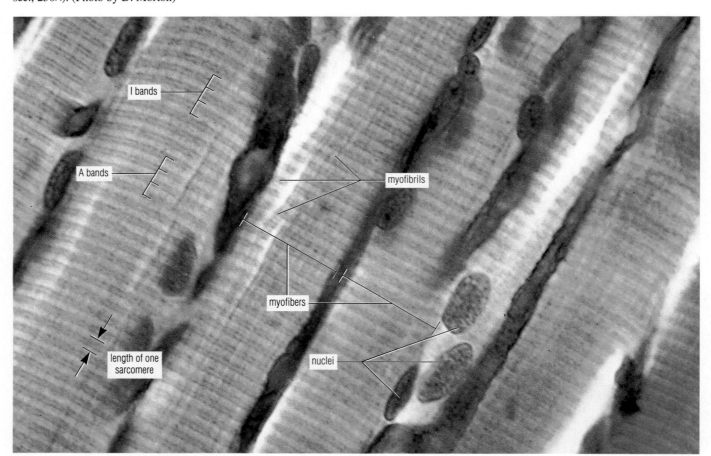

Figure 20c Silver-stained **skeletal muscle cells** (prep. slide, l.s., 1800×). (Photo by D. Morton)

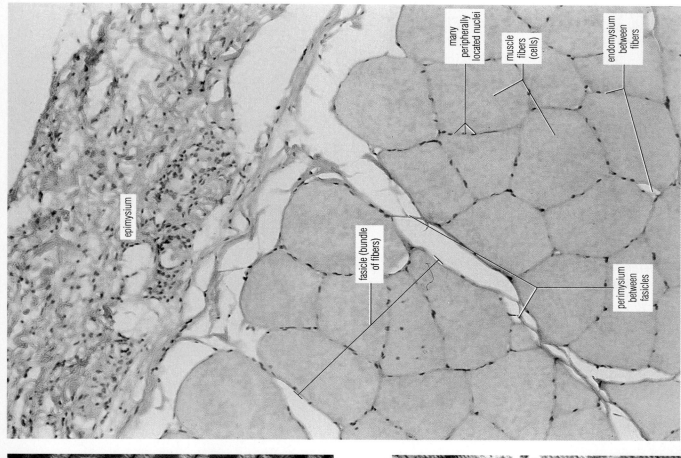

Figure 21c Skeletal muscle organ (c.s., 150×). (Photo by D. Morton)

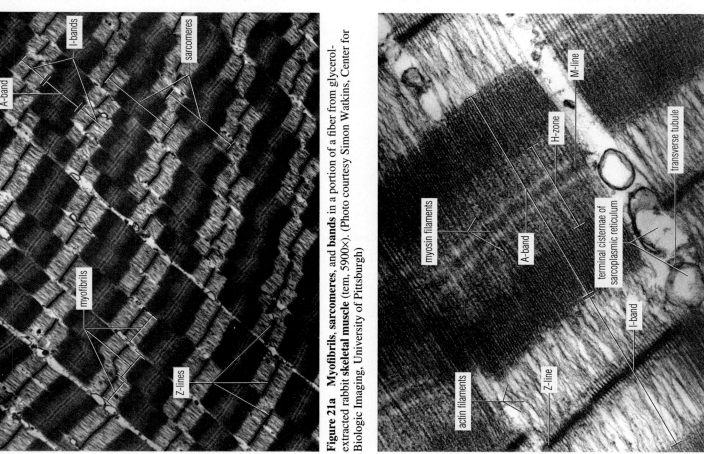

Figure 21a Myofibrils, sarcomeres, and bands in a portion of a fiber from glycerol-extracted rabbit skeletal muscle (tem, 5900×). (Photo courtesy Simon Watkins, Center for Biologic Imaging, University of Pittsburgh)

Figure 21b Actin and myosin myofilaments in a sarcomere from glycerol-extracted rabbit skeletal muscle (tem, 29,500×). (Photo courtesy Simon Watkins, Center for Biologic Imaging, University of Pittsburgh)

22 TISSUES Muscle Tissue/Cardiac and Smooth

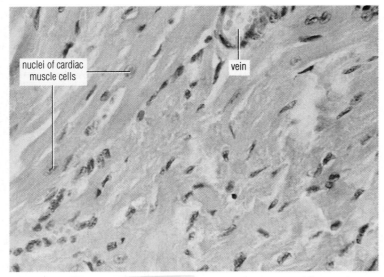

Figure 22a Cardiac muscle tissue in a section of the heart (prep. slide, 100×). (Photo by D. Morton)

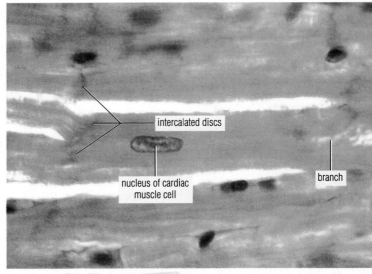

Figure 22b Cardiac muscle cells. Striations are present but faint (prep. slide, l.s., 1000×). (Photo by D. Morton)

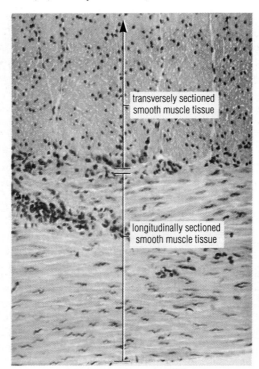

Figure 22c Smooth muscle tissue in **muscularis externa** of **small intestine** (prep. slide, l.s., 250×). (Photo by D. Morton)

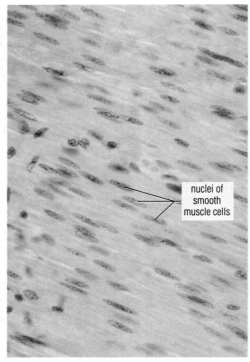

Figure 22d Longitudinal section of **smooth muscle cells** (prep. slide, 400×). (Photo by D. Morton)

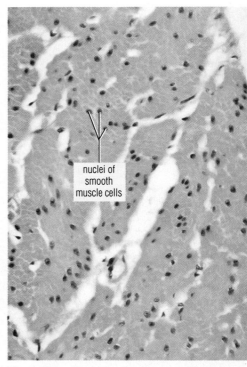

Figure 22e Cross section of **smooth muscle cells.** Not all of the smooth muscle cells show nuclei, because in many fibers they are in the previous or subsequent sections (prep. slide, 400×). (Photo by D. Morton)

Nervous Tissue/Spinal Cord **TISSUES** 23

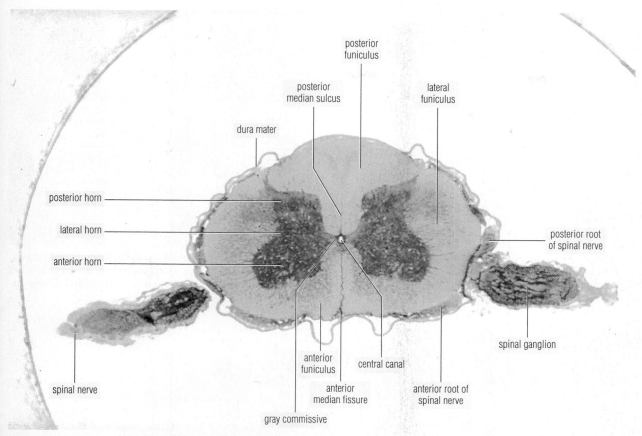

Figure 23a Cross section of spinal cord at thoracic level (prep. slide, 10×). (Photo by D. Morton)

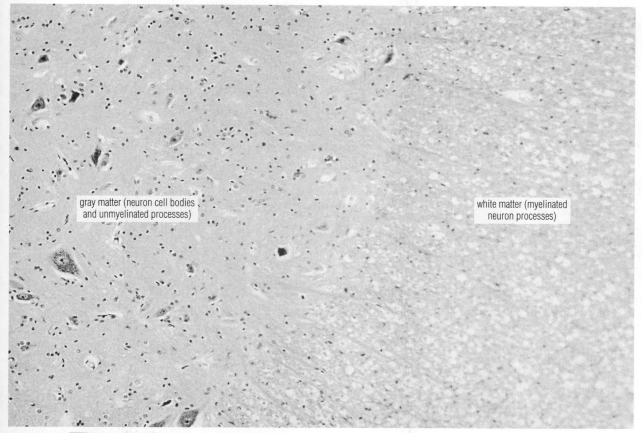

Figure 23b Nervous tissue—gray matter and white matter—of spinal cord (prep. slide, c.s., 100×). (Photo by D. Morton)

Nervous Tissue/Spinal Cord

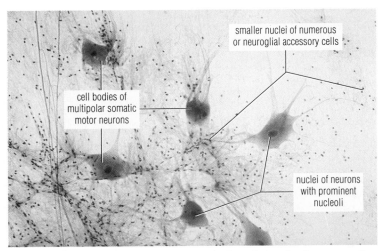

Figure 24a **Smear** of **gray matter** of **spinal cord** showing large **somatic motor neurons**, and associated **neuroglia** (accessory cells). These neurons are referred to as multipolar because they have multiple processes, including one axon and many dendrites (prep. slide, w.m., 100×). (Photo by D. Morton)

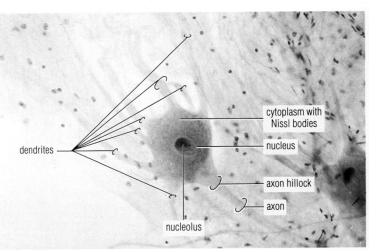

Figure 24b Higher-power view of **somatic motor neuron**. **Nissl bodies** are concentrations of rough endoplasmic reticulum and are not present in the root of the axon called the **axon hillock** (prep. slide, w.m., 250×). (Photo by D. Morton)

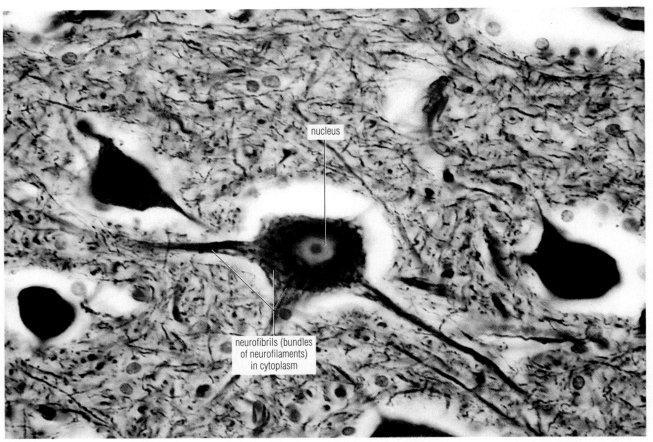

Figure 24c Silver-stained **somatic motor neuron** in the ventral horn of the spinal cord showing neurofibrils, or bundles of **neurofilaments**, in the cytoplasm (prep. slide, c.s., 650×). (Photo by D. Morton)

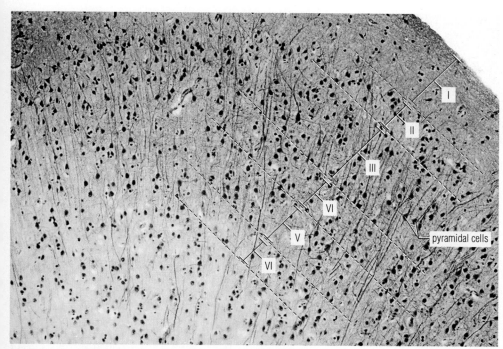

Figure 25a Silver-stained section exhibiting the **six layers** of **cerebral cortex**: molecular layer (I), outer granular layer with small pyramidal-shaped cells (II), outer pyramidal cell layer with medium-size pyramidal cells (III), inner granular layer (IV), inner pyramidal cell layer with small pyramidal cells (V), and fusiform layer (VI). In the motor region of the cerebral cortex, large pyramidal cells called Betz cells are present in the fifth layer, which there is called the ganglionic layer (prep. slide, sec., 30×). (Photo by D. Morton)

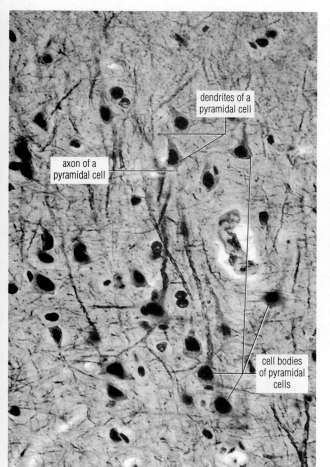

Figure 25b Higher-power view of **pyramidal cells** (neurons) in the third layer of the **cerebral cortex** (prep. slide, sec., 300×). (Photo by D. Morton)

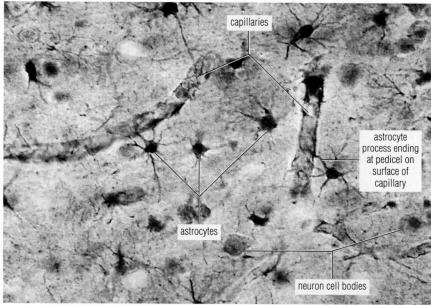

Figure 25c **Astrocytes**—one type of neuroglial cell—in a silver-stained section of a part of the central nervous system. Their processes end in **pedicels** (foot processes) that form a discontinuous membrane over capillaries and other structures (prep. slide, 500×). (Photo by D. Morton)

26 TISSUES *Nervous Tissue/Brain/Cerebellum*

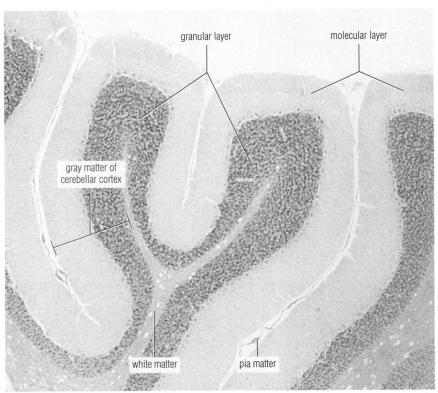

Figure 26a Folds in the surface of the **cerebellum** (prep. slide, sec., 40×). (Photo by D. Morton)

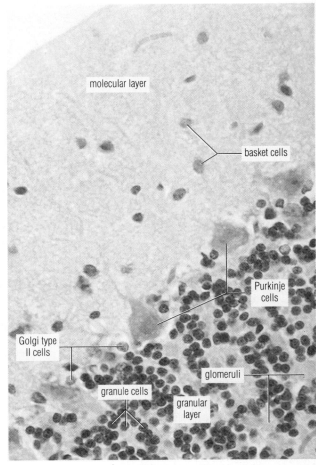

Figure 26b Gray matter of **cortex** of the **cerebellum** (prep. slide, sec., 300×). (Photo by D. Morton)

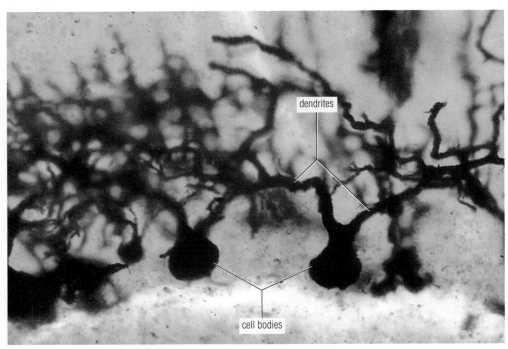

Figure 26c Purkinje cells (neurons) in a silver-stained thick section of the **cerebellar cortex** (prep. slide, 550×). (Photo by D. Morton)

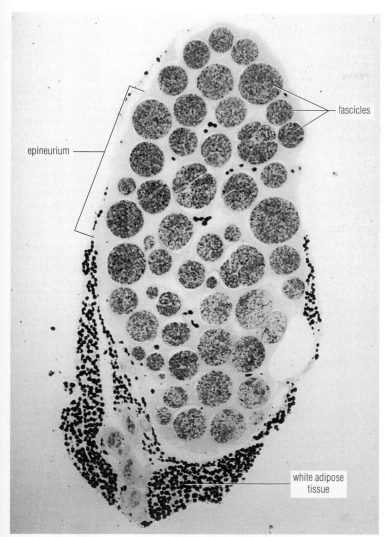

Figure 27a Cross section of **sciatic nerve** stained with osmium tetroxide, which fixes lipids and stains them black. The **epineurium** is the dense, irregular connective tissue surrounding the entire nerve (prep. slide, 35×). (Photo by D. Morton)

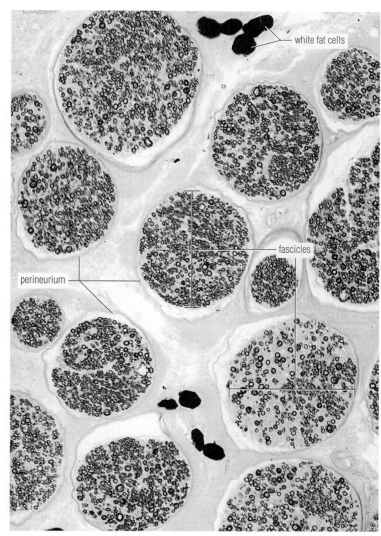

Figure 27b Higher-power view of the interior of the sciatic nerve cross section. The **perineurium** surrounds each **fascicle** and is composed of irregular connective tissue slightly less dense than that of the epineurium (prep. slide, c.s., 150×). (Photo by D. Morton)

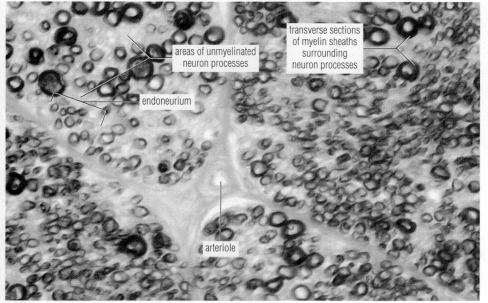

Figure 27c Higher-power view of the interior of a **fascicle** in the sciatic nerve cross section, which contains both myelinated and unmyelinated neuron processes surrounded by Schwann cells. Only the **myelin sheaths** are deeply stained. A segment of the myelin sheath is formed by a Schwann cell wrapping itself many times around a neuron process while squeezing out the cytoplasm from between the two sides of its plasma membrane. The **endoneurium** is loose connective tissue situated between the outer membranes of adjacent Schwann cells (prep. slide, c.s., 500×). (Photo by D. Morton)

28 TISSUES Nervous Tissue/Nerves/Motor Units

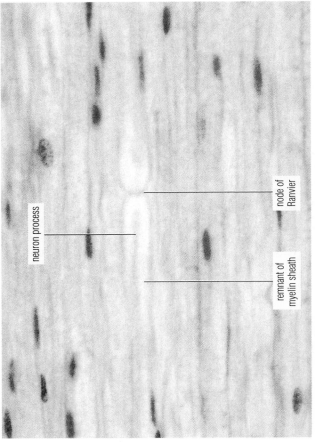

Figure 28b **Longitudinal section** of a **nerve** showing myelinated neuron processes. **Nodes of Ranvier** are interruptions of the **myelin sheath** where one Schwann cell segment ends and the next begins (prep. slide, 650×). (Photo by D. Morton)

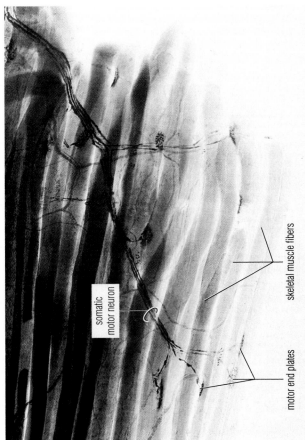

Figure 28d Silver-stained skeletal muscle tissue showing portions of **motor units**, each of which consists of a somatic motor neuron and the fibers it innervates (prep. slide, w.m., 100×). (Photo by D. Morton)

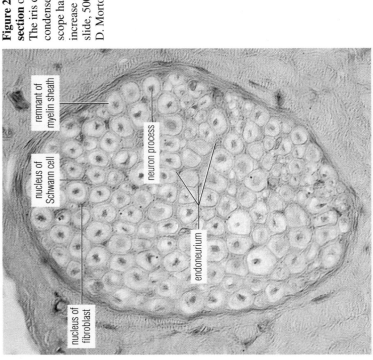

Figure 28a **Cross section** of a small **nerve**. The iris diaphragm of the microscope condenser has been closed to increase contrast (prep. slide, 500×). (Photo by D. Morton)

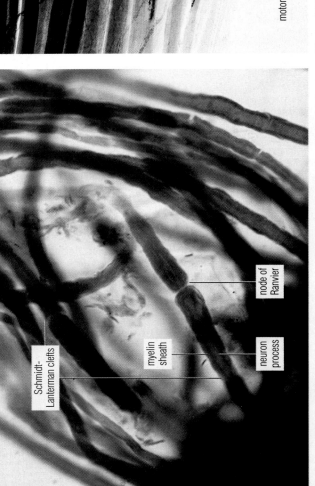

Figure 28c **Teased nerve** stained with osmium tetroxide. The **myelin sheaths** are stained black. A **Schmidt-Lanterman cleft** is an area where Schwann cell cytoplasm was not squeezed out during formation of the sheath segment (prep. slide, w.m., 500×). (Photo by D. Morton)

Nervous Tissue/Ganglia/Sensory Structures — TISSUES

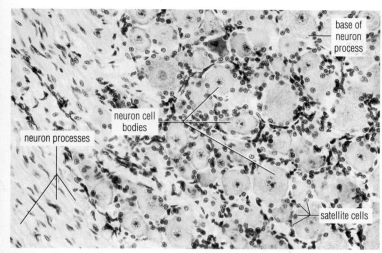

Figure 29a **Spinal ganglion**, or dorsal root ganglion, containing the cell bodies of **somatic sensory neurons** and **visceral sensory neurons**. These neurons are referred to as "pseudounipolar" because their single process has both afferent and efferent tracts. Satellite cells tend to be evenly distributed around the roundish cell bodies (prep. slide, sec., 250×). (Photo by D. Morton)

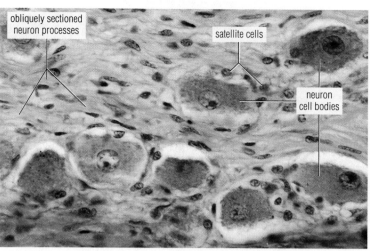

Figure 29b **Sympathetic ganglion**, containing postganglionic **visceral motor neurons**. Their cell bodies are more angular in shape due to the numerous processes of these multipolar neurons—processes that disrupt the placement of the surrounding satellite cells (prep. slide, sec., 400×). (Photo by D. Morton)

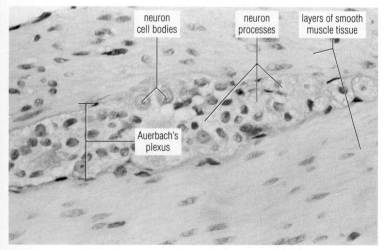

Figure 29c **Auerbach's plexus** located between the smooth muscle layers of the musclularis externa in the digestive tract (prep. slide, sec., 400×). (Photo by D. Morton)

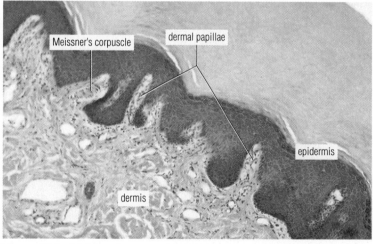

Figure 29d **Meissner's corpuscles** or touch receptors, a type of encapsulated neuron ending located in the dermal papillae of hairless skin, such as the lips and finger-tips (prep. slide, sec., 100×). (Photo by D. Morton)

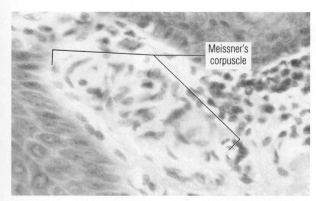

Figure 29e High-power view of a **Meissner's corpuscle**, consisting of one or two neuron endings (not easily seen) coursing through a capsule of supportive cells, which are arranged perpendicularly to its longitudinal axis (prep. slide, sec., 500×). (Photo by D. Morton)

Figure 29f **Pacinian corpuscle**, an encapsulated neuron ending sensitive to pressure. It is located in the subcutaneous layer (hypodermis) of the skin and other locations (prep. slide, sec., 100×). (Photo by D. Morton)

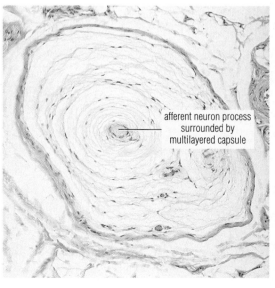

INTEGUMENTARY SYSTEM *Skin*

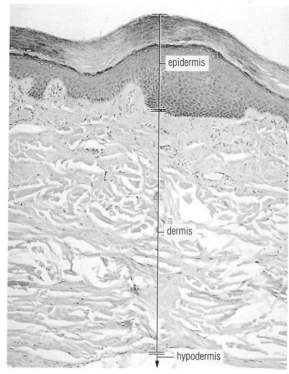

Figure 30a **Thick skin** from the palm of the hand (prep. slide, sec., 30×). (Photo by D. Morton)

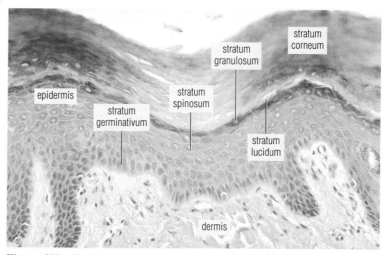

Figure 30b **Epidermal strata** in a section of thick skin. The strata result from the process of **keratinization**, or synthesis of the protein keratin. During this process, new keratinocytes (the predominant cell type in the epidermis) produced by cell division in the basal stratum move toward the surface to be shed after a journey of a few weeks. In individuals with darker skin tones, melanin pigment will be seen in the keratinocytes of the stratum germinativum. Melanin is synthesized by melanocytes (not seen) and passed on to the keratinocytes (prep. slide, 100×). (Photo by D. Morton)

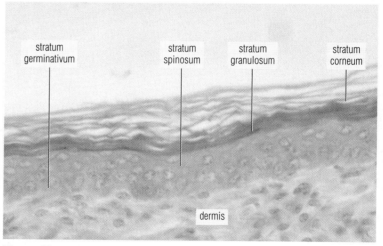

Figure 30c **Thin skin** (prep. slide, sec., 250×). (Photo by D. Morton)

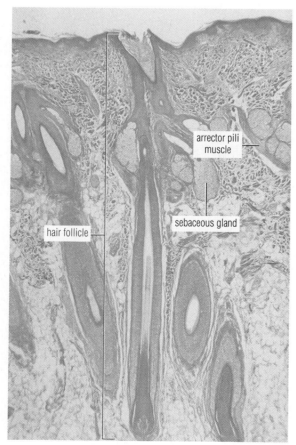

Figure 30d **Hairy thin skin** with **longitudinally sectioned hair follicle** (prep. slide, 8×). (Photo by D. Morton)

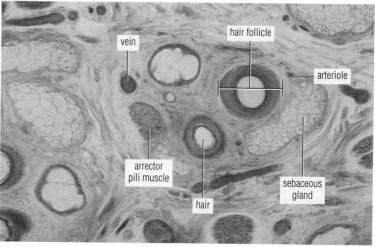

Figure 30e **Hairy thin skin**, stained with a trichrome procedure, with **cross-sectioned hair follicles** (prep. slide, 100×). (Photo by D. Morton)

Skin **INTEGUMENTARY SYSTEM** 31

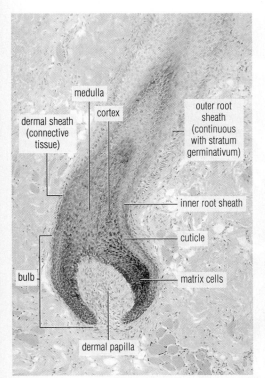

Figure 31a **Bulb** of **hair follicle**. The **matrix** contains dividing keratinocytes and pigment-producing melanocytes (not seen) (prep. slide, sec., 100×). (Photo by D. Morton)

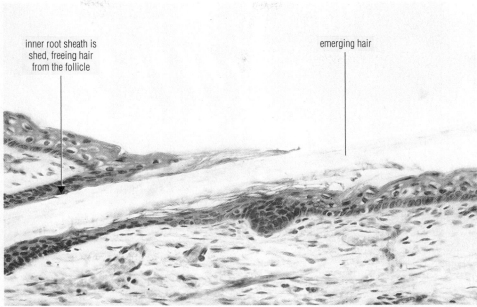

Figure 31c **Hair release** from the sides of the follicles happens when keratinocytes of the inner root sheath are shed just below the level of the sebaceous gland ducts (prep. slide, sec., 100×). (Photo by D. Morton)

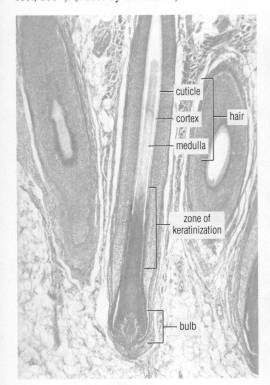

Figure 31b **Hair formation** occurs as keratinocytes of the medulla, cortex, and cuticle synthesize keratin. Unlike the soft keratin of epidermal keratinocytes, structures made of hard keratin are shed in one piece (hair) or in pieces (nails) (prep. slide, sec., 20×). (Photo by D. Morton)

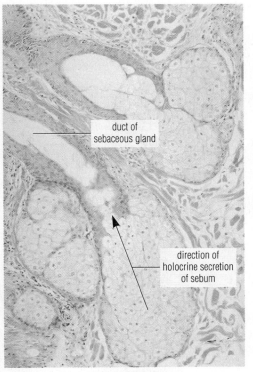

Figure 31d The **sebaceous gland** secretes **sebum** by a holocrine mechanism when the cells nearest the ducts burst to release their contents. This waxy substance lubricates the hair and skin surface (prep. slide, sec., 100×). (Photo by D. Morton)

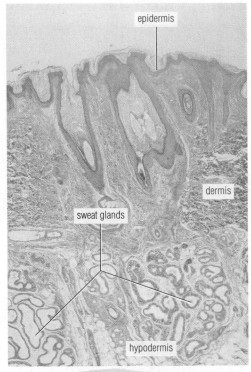

Figure 31e **Sweat glands** (prep. slide, sec., 20×). (Photo by D. Morton)

Skeletal System

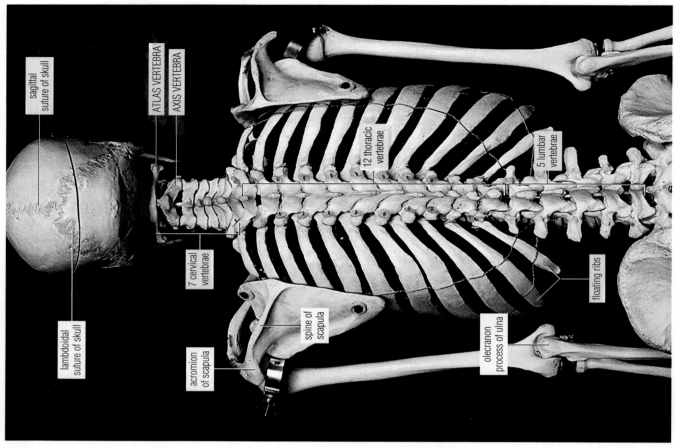

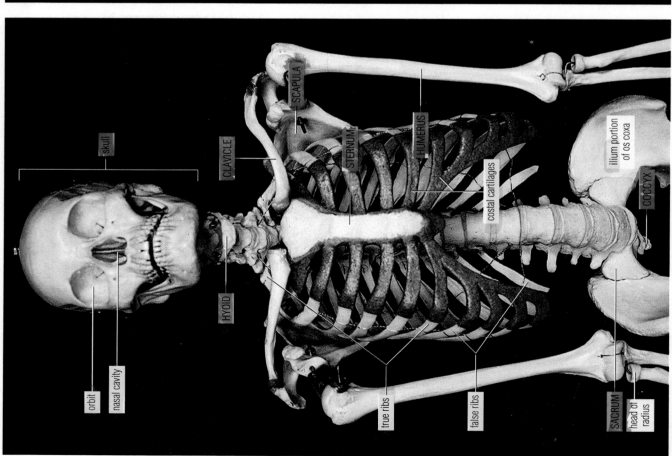

Figure 32 **Anterior** (left) and **posterior** (right) views of the upper two-thirds of the **human skeleton**. The names of bones are in capital letters, while features (markings) of a bone and structures made up of two or more bones are lowercase (0.30×). (Photos by D. Morton)

Skeletal System — Skull

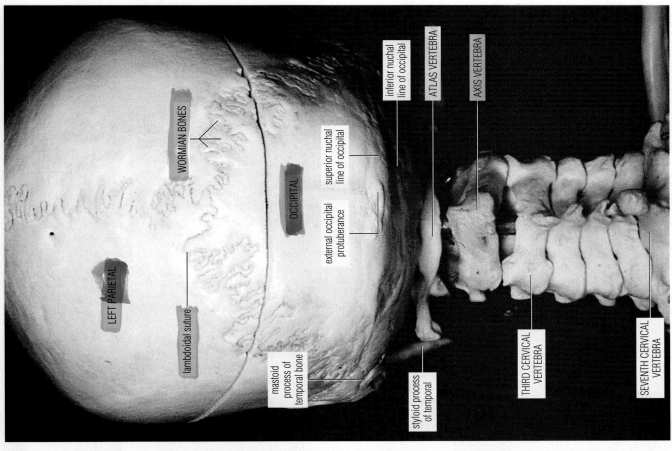

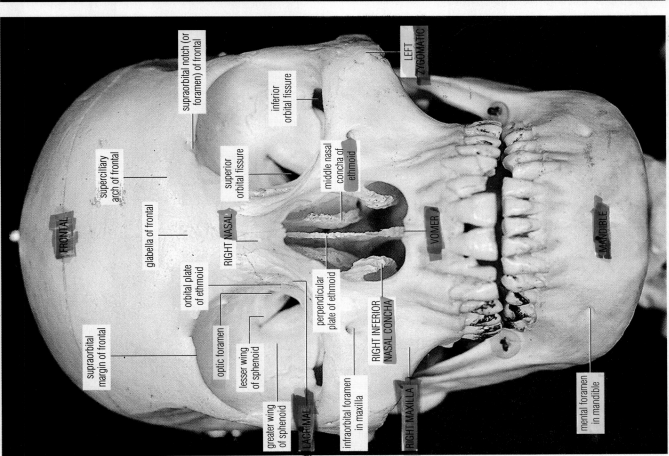

Figure 33 **Anterior** (left) and **posterior** (right) views of the **skull**. The bones of the skull are divided into two groups. One group—frontal, parietal, temporal, sphenoid, and ethmoid—forms the **cranium**. The other group—maxilla, palatine (internal and therefore not visible; see Figure 34b), zygomatic, lacrimal, nasal, inferior nasal concha, vomer, and mandible—constitutes the **facial bones** (1×). (Photos by D. Morton)

34 Skeletal System — Skull

Figure 34a Lateral view of left side of skull (0.50×). (Photo by D. Morton)

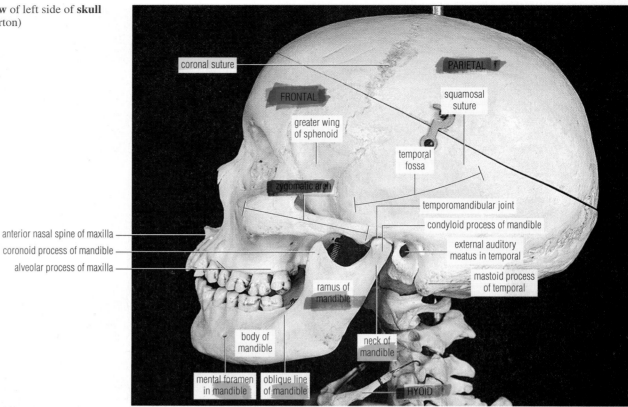

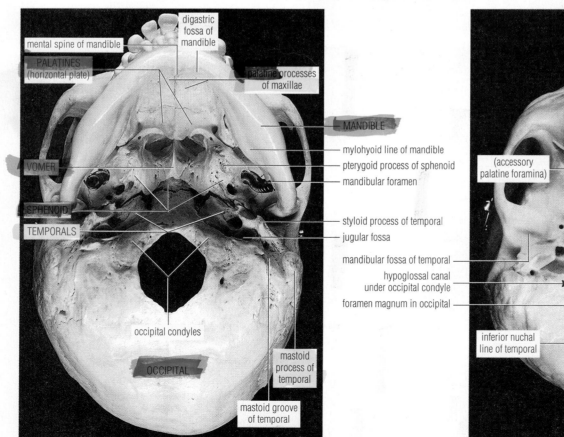

Figure 34b Inferior view of skull (natural bone). The palatine processes of the maxillae and the horizontal plate of the palatines form the **hard palate** (0.70×). (Photo by D. Morton)

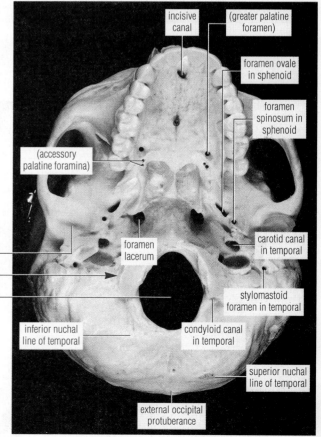

Figure 34c Inferior view of skull (plastic) without the mandible (0.70×). (Photo by D. Morton)

Skull **SKELETAL SYSTEM** 35

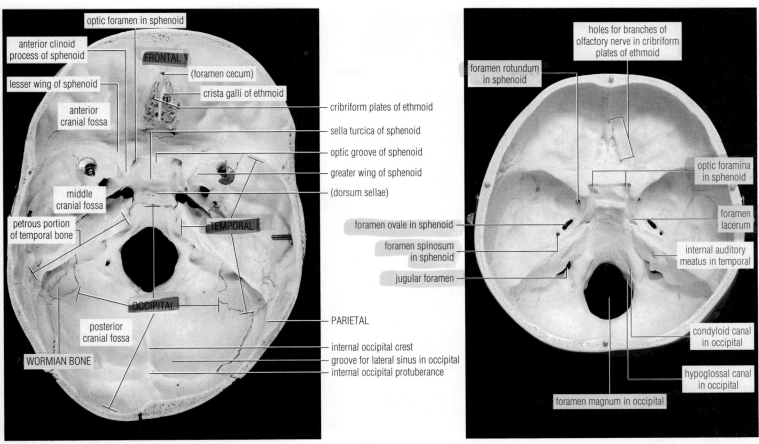

Figure 35a The **floor** of the **cranial cavity** (natural-bone skull) (0.70×). (Photo by D. Morton)

Figure 35b The **floor** of the **cranial cavity** (plastic skull) (0.50×). (Photo by D. Morton)

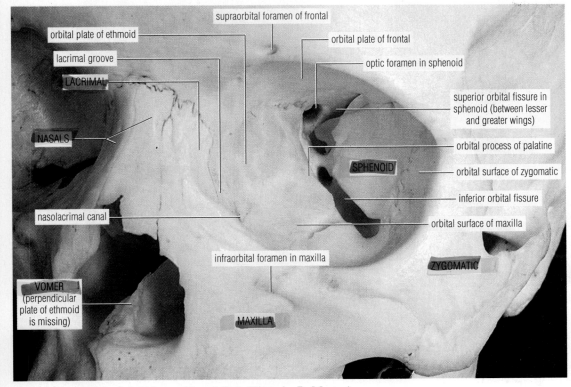

Figure 35c **Orbit** of skull (natural bone) (2×). (Photo by D. Morton)

36 Skeletal System *Skull*

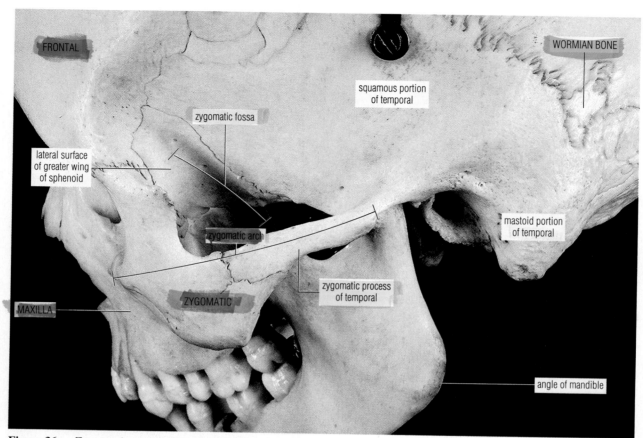

Figure 36a **Zygomatic arch** and **fossa** of skull (natural bone) (1×). (Photo by D. Morton)

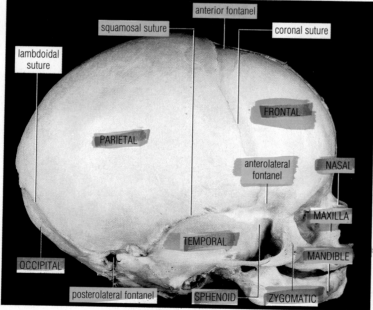

Figure 36b **Lateral view** of **fetal skull** (natural bone) (0.90×). (Photo by D. Morton)

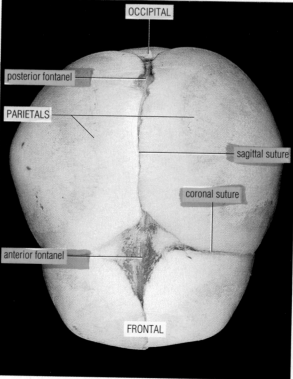

Figure 36c **Superior view** of **fetal skull** (natural bone) (0.90×). (Photo by D. Morton)

Skeletal System

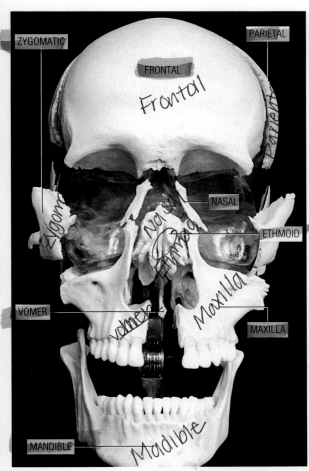

Figure 37a Anterior view of **disarticulated skull.** In this particular preparation the palatine, lacrimal, and inferior nasal concha paired bones are attached to the maxillae (plastic) (0.50×). (Photo by D. Morton)

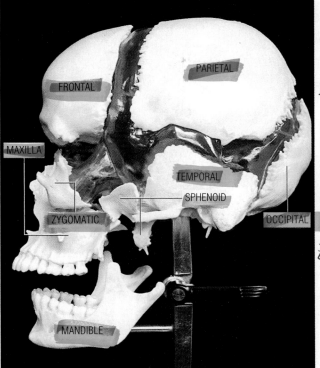

Figure 37b Lateral view of left side of **disarticulated skull** (plastic) (0.40×). (Photo by D. Morton)

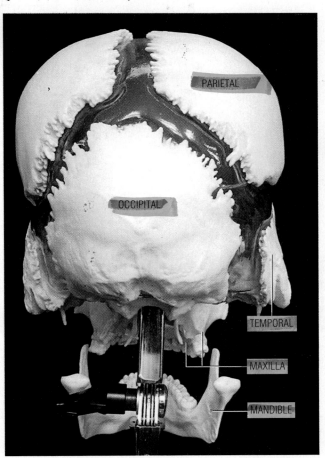

Figure 37c Posterior view of disarticulated skull (plastic) (0.50×). (Photo by D. Morton)

Handwritten annotations:

On Figure 37a: Frontal, Parietal, Zygom, Nasal, Ethmoid, vomer, Maxilla, Madible

Side - parietal
frontal - front
Temporal - side lower
maxilla - above teeth
zygomatic - cheek
madible - jaw

sagittal suture - line down middle head
anterior frontanel - middle indend. on top
coronal suture - side lines
posterior fontanel - indent in back.

occipital - back head

38 SKELETAL SYSTEM Skull Bones

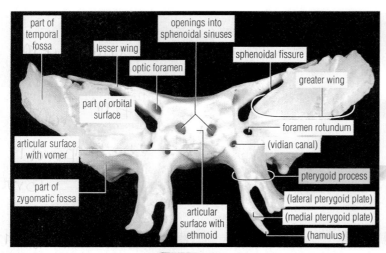

Figure 38a Anterior view of **sphenoid** bone (plastic) (0.70×). (Photo by D. Morton)

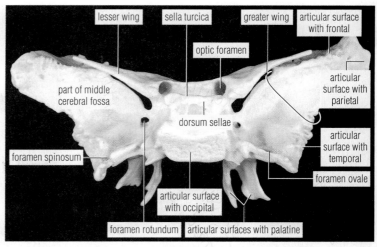

Figure 38b Posterior view of **sphenoid** bone (plastic) (0.70×). (Photo by D. Morton)

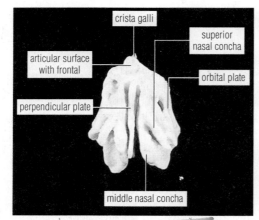

Figure 38c Anterior view of **ethmoid** bone (plastic) (0.80×). (Photo by D. Morton)

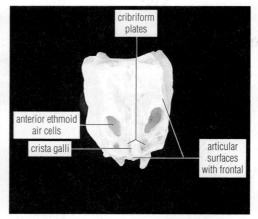

Figure 38d Superior view of **ethmoid** bone (plastic) (0.80×). (Photo by D. Morton)

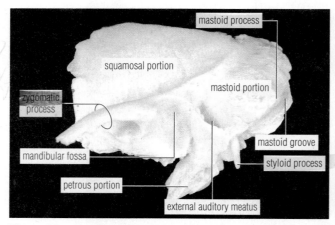

Figure 38e Lateral view of external surface of **temporal** bone (plastic) (0.70×). (Photo by D. Morton)

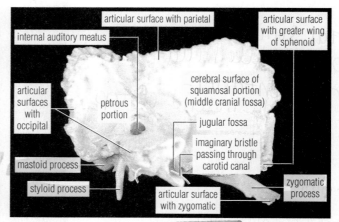

Figure 38f Cranial surface of **temporal** bone (plastic) (0.70×). (Photo by D. Morton)

Skull Bones **SKELETAL SYSTEM** 39

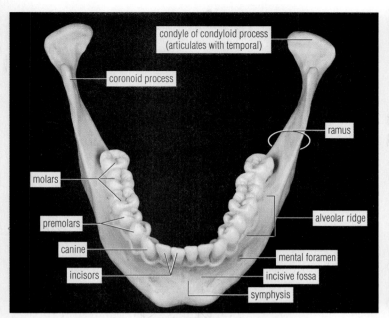

Figure 39a Superior view of **mandible** (plastic) bone (0.70×). (Photo by D. Morton)

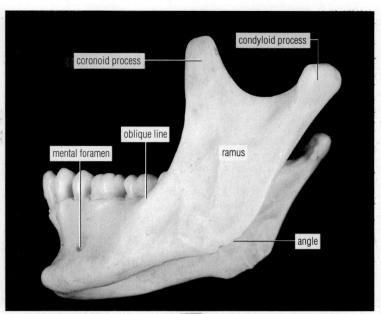

Figure 39b Lateral view of **mandible** (plastic) (0.90×). (Photo by D. Morton)

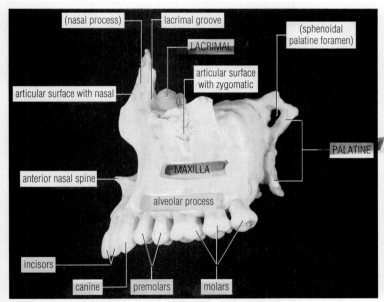

Figure 39c Lateral view of left **maxilla**, lacrimal, and palatine bones (plastic) (0.80×). (Photo by D. Morton)

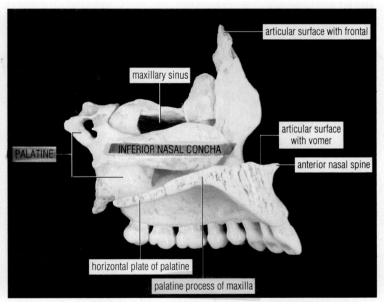

Figure 39d Medial view of left **maxilla**, palatine, and inferior nasal concha bones (plastic) (0.90×). (Photo by D. Morton)

Skeletal System — Vertebrae

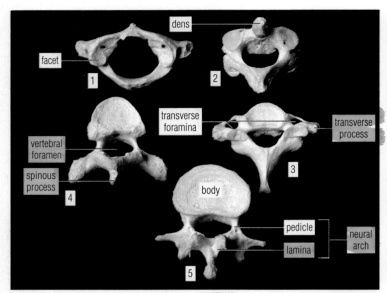

Figure 40a Types of vertebrae: (1) atlas (first cervical vertebra), (2) axis (second cervical vertebra), (3) typical of the rest of the cervical vertebrae, (4) thoracic vertebra, (5) lumbar vertebra (0.40×). (Photo by D. Morton)

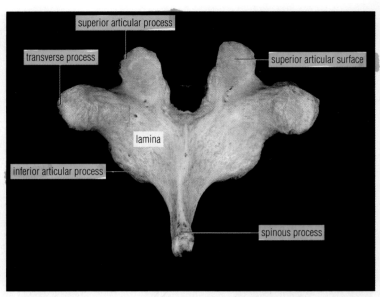

Figure 40b Posterior view of a thoracic vertebra (1×). (Photo by D. Morton)

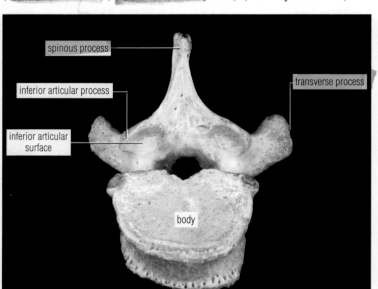

Figure 40c Inferior view of a thoracic vertebra (1×). (Photo by D. Morton)

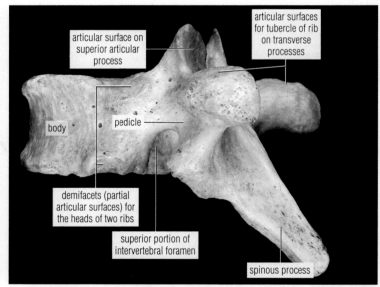

Figure 40d Lateral view of a thoracic vertebra (120×). (Photo by D. Morton)

Sacrum/Sternum/Rib — SKELETAL SYSTEM

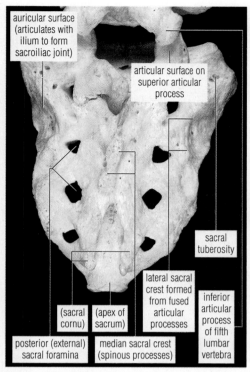

Figure 41a Posterior view of sacrum (0.70×). (Photo by D. Morton)

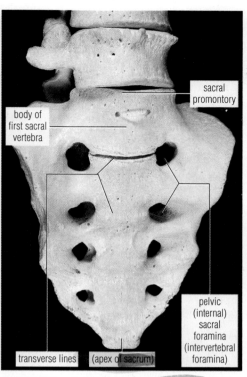

Figure 41b Anterior view of sacrum (0.70×). (Photo by D. Morton)

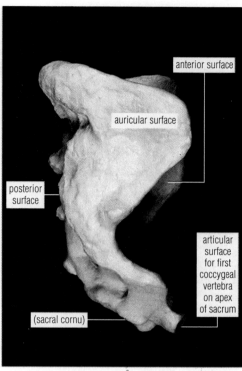

Figure 41c Lateral view of sacrum (0.50×). (Photo by D. Morton)

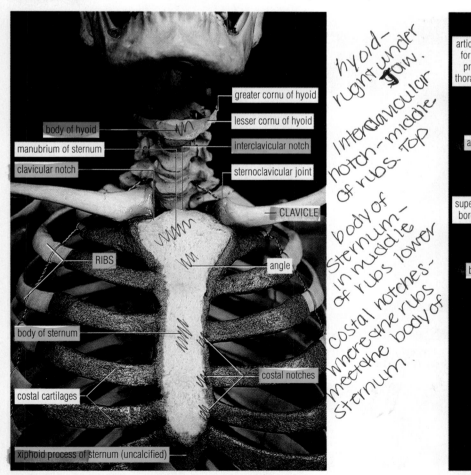

Figure 41d Anterior view of hyoid and sternum (0.60×). (Photo by D. Morton)

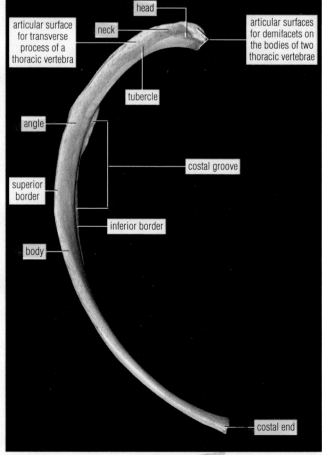

Figure 41e Superior view of a right rib (0.70×). (Photo by D. Morton)

42 SKELETAL SYSTEM *Pectoral Girdle/Bones of Shoulder Joint/Scapula/Clavicle*

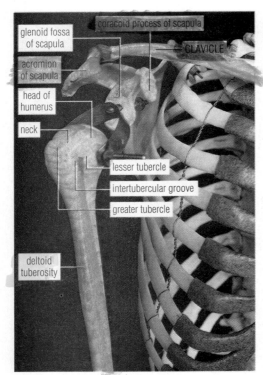

Figure 42a Anterior view of the **bones** of the right **shoulder joint** (0.50×). (Photo by D. Morton)

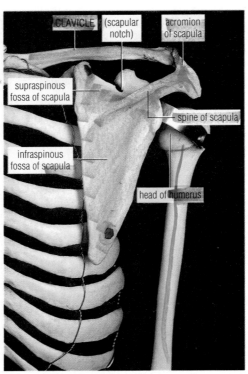

Figure 42b Posterior view of the **bones** of the right **shoulder joint** (0.40×). (Photo by D. Morton)

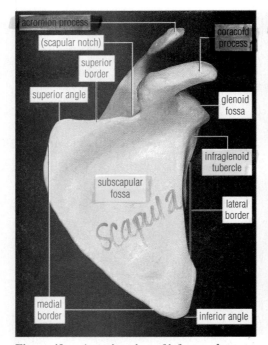

Figure 42c Anterior view of left **scapula** (plastic) (0.50×). (Photo by D. Morton)

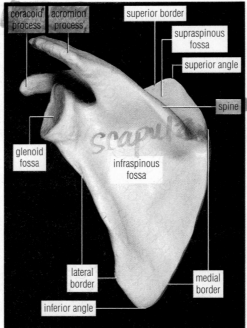

Figure 42d Posterior view of left **scapula** (plastic) (0.50×). (Photo by D. Morton)

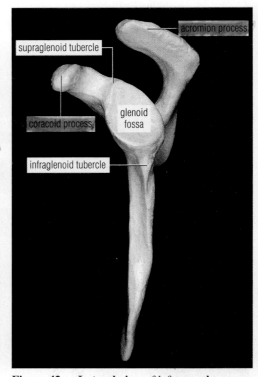

Figure 42e Lateral view of left **scapula** (plastic) (0.50×). (Photo by D. Morton)

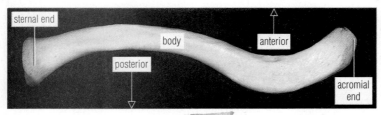

Figure 42f Superior view of right **clavicle** (0.70×). (Photo by D. Morton)

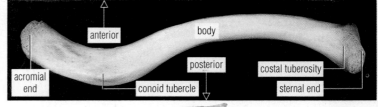

Figure 42g Inferior view of right **clavicle** (0.70×). (Photo by D. Morton)

Bones of Elbow and Wrist Joints/Bones of Hand — Skeletal System

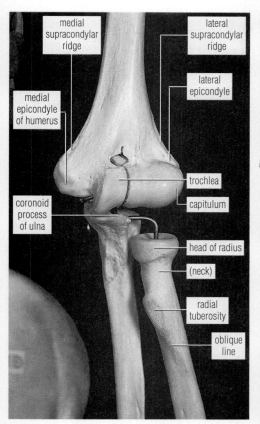

Figure 43a Anterior view of the bones of the left elbow joint (0.75×). (Photo by D. Morton)

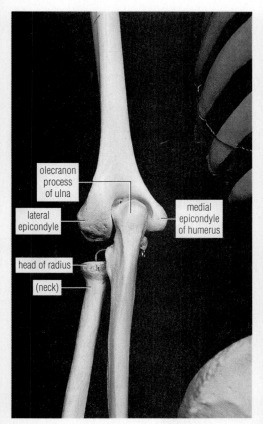

Figure 43b Posterior view of the bones of the left elbow joint (0.50×). (Photo by D. Morton)

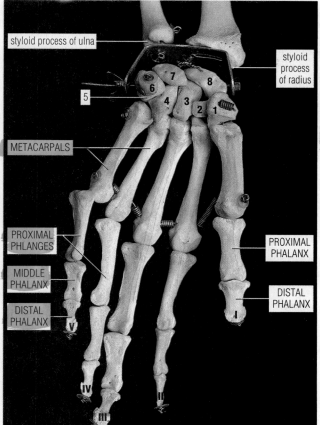

Figure 43c Anterior view of the bones of the left wrist joint and hand. The eight carpal bones are numbered:
(1) trapezium
(2) trapezoid
(3) capitate
(4) hamate
(5) triquetrum
(6) pisiform
(7) lunate
(8) scaphoid
(0.60×). (Photo by D. Morton)

Carpals 11

1 starts over the thumb
2,3,4,5
6 on top of 5
7 above 4
8 above 3,2,1

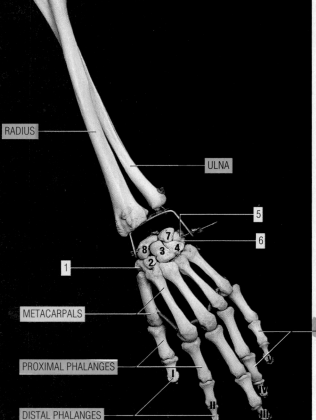

Figure 43d Posterior view of left hand with the forearm pronated. Carpal bones are numbered as in Figure 43c (0.40×). (Photo by D. Morton)

Skeletal System — Bones of Arm

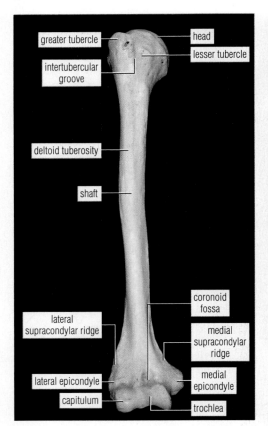

Figure 44a Anterior view of right **humerus** (0.40×). (Photo by D. Morton)

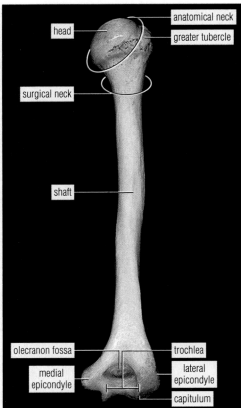

Figure 44b Posterior view of right **humerus** (0.40×). (Photo by D. Morton)

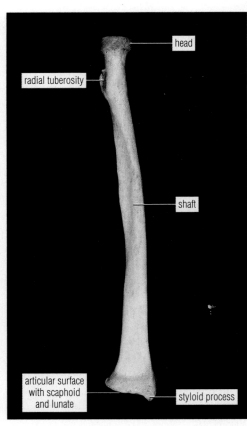

Figure 44c Anterior view of left **radius** (0.40×). (Photo by D. Morton)

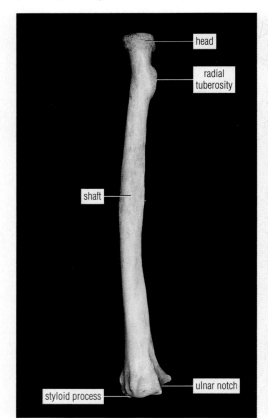

Figure 44d Posterior view of left **radius** (0.40×). (Photo by D. Morton)

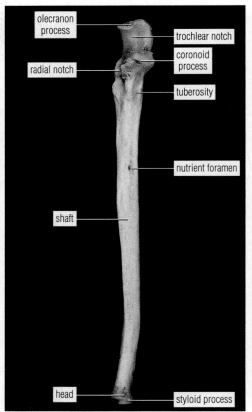

Figure 44e Anterior view of right **ulna** (0.40×). (Photo by D. Morton)

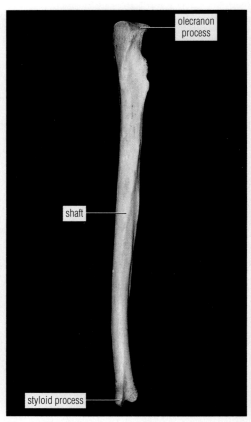

Figure 44f Posterior view of right **ulna** (0.40×). (Photo by D. Morton)

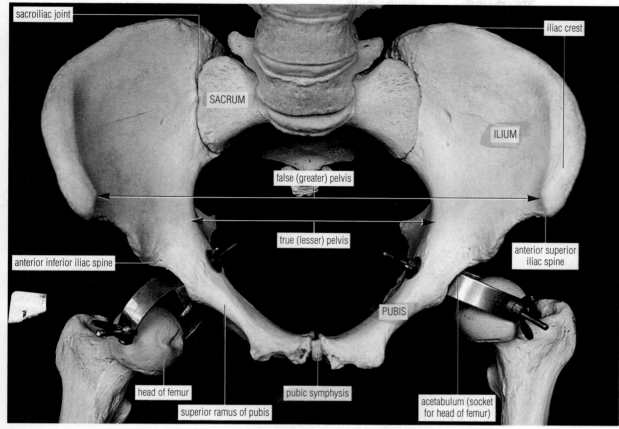

Figure 45a Anterior view of the bones of the female pelvis and hip joint (0.70×). (Photo by D. Morton)

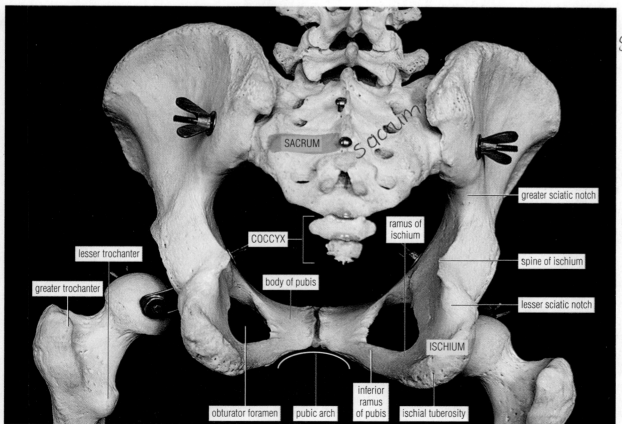

Figure 45b Posterior view of the bones of the female pelvis and hip joint (0.70×). (Photo by D. Morton)

Skeletal System — Male Pelvic Girdle

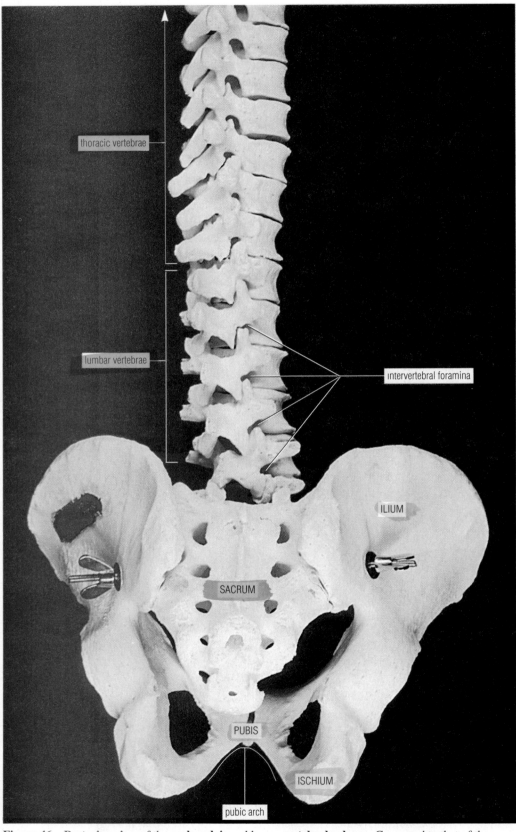

cervical vertebrae - 7
thoracic vertebra - 12
lumbar vertebra 5
sacrum - 4 to 5
coccyx - 4 to 5

Figure 46 **Posterior view** of the **male pelvis** and lower **vertebral column**. Compared to that of the female, the male pelvis is relatively heavier with more prominent processes, its inlet is more heart-shaped, and its outlet is narrower. Also, the pubic arch of the male pelvis is more acute (less than 90°) and the tip of the coccyx (not shown) tilts more anteriorly than in the female (0.50×). (Photo by D. Morton)

Os Coxa SKELETAL SYSTEM 47

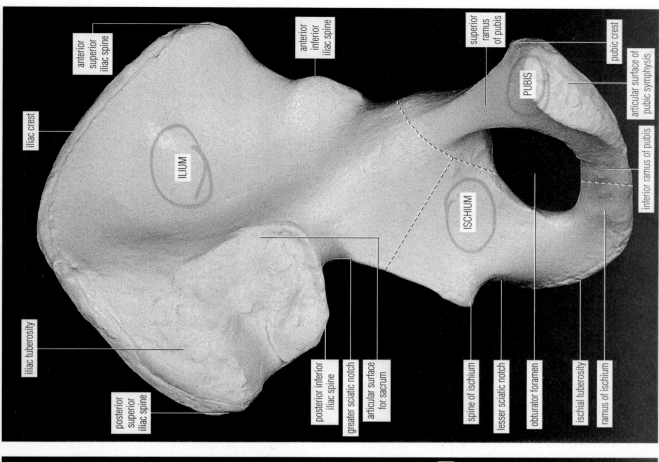

Figure 47b Medial view of the left os coxa (plastic) (1×). (Photo by D. Morton)

Figure 47a Lateral view of the left os coxa (plastic) (1×). (Photo by D. Morton)

general term for the hips.

Skeletal System — Bones of Knee/Femur/Tibia

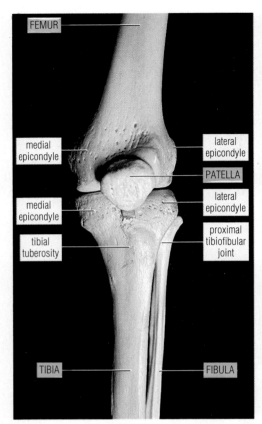

Figure 48a Anterior view of the bones of the left **knee joint** (0.40×). (Photo by D. Morton)

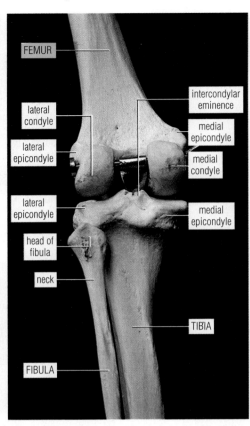

Figure 48b Posterior view of the bones of the left **knee joint** (0.50×). (Photo by D. Morton)

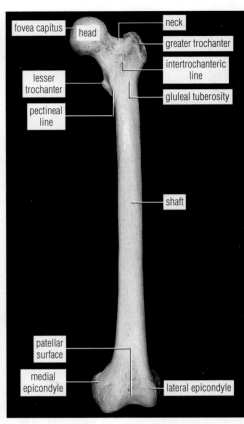

Figure 48c Anterior view of left **femur** (0.30×). (Photo by D. Morton)

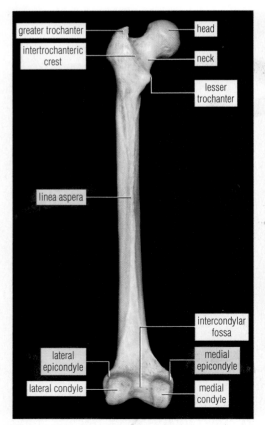

Figure 48d Posterior view of left **femur** (0.30×). (Photo by D. Morton)

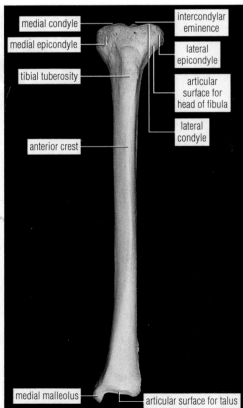

Figure 48e Anterior view of left **tibia** (0.30×). (Photo by D. Morton)

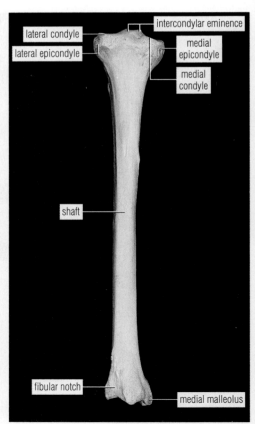

Figure 48f Posterior view of left **tibia** (0.30×). (Photo by D. Morton)

Fibula/Bones of Foot — SKELETAL SYSTEM

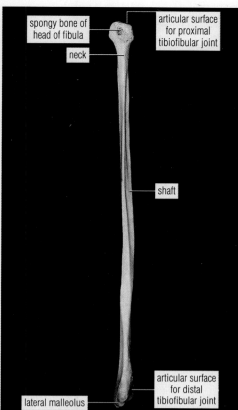

Figure 49a Anterior view of right **fibula** (0.30×). (Photo by D. Morton)

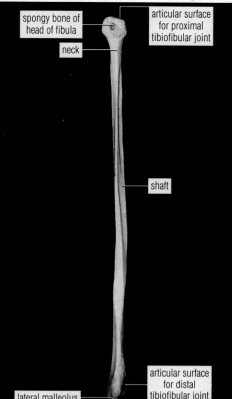

Figure 49c Lateral view of the **bones** of the left **ankle joint** and **foot**. The seven tarsal bones are numbered:

(1) talus
(2) first cuneiform
(3) second cuneiform
(4) third cuneiform
(5) cuboid
(6) naviculus
(7) calcaneus

(0.60×). (Photo by D. Morton)

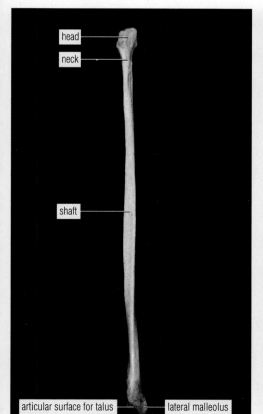

Figure 49b Posterior view of left **fibula** (0.30×). (Photo by D. Morton)

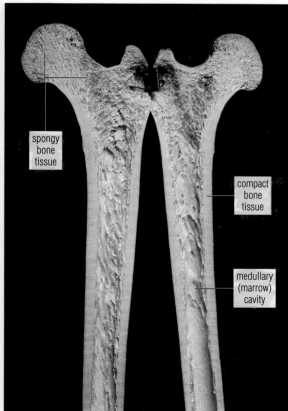

Figure 49d Sawed **femur** (0.40×). (Photo by D. Morton)

50 ARTICULATIONS *Shoulder and Elbow*

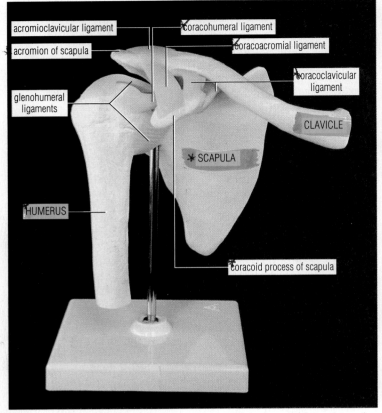

Figure 50a Anterior view of a model of the shoulder joint (0.45×). (Photo by D. Morton)

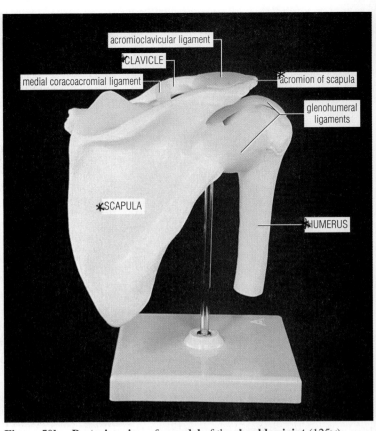

Figure 50b Posterior view of a model of the shoulder joint (125×). (Photo by D. Morton)

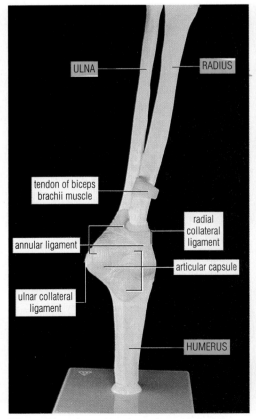

Figure 50c Anterior view of a model of the elbow joint (0.30×). (Photo by D. Morton)

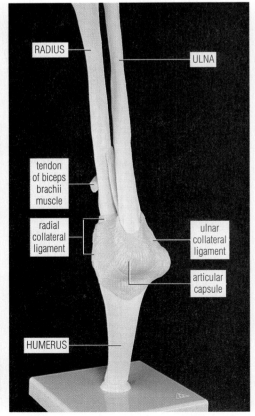

Figure 50d Posterior view of a model of the elbow joint (0.30×). (Photo by D. Morton)

Hip/Developing Joint   **ARTICULATIONS**   51

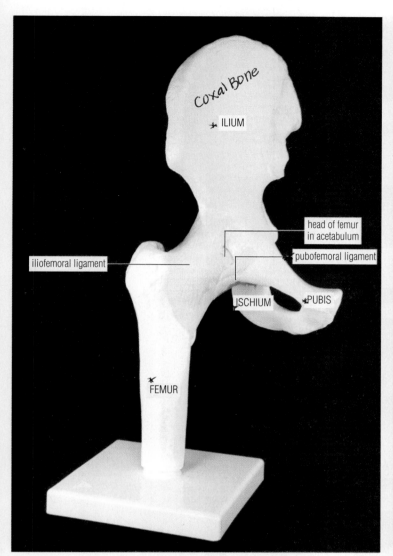

Figure 51a **Anterior view** of a **model** of the **hip joint** (0.40×). (Photo by D. Morton)

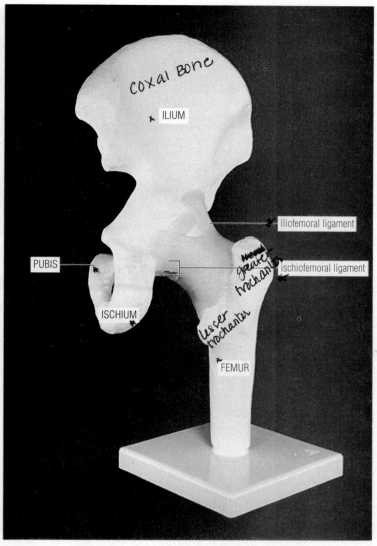

Figure 51b **Posterior view** of a **model** of the **hip joint** (0.40×). (Photo by D. Morton)

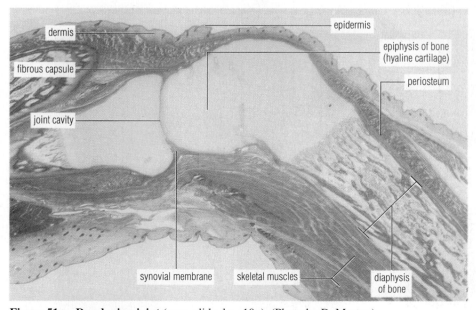

Figure 51c **Developing joint** (prep. slide, l.s., 10×). (Photo by D. Morton)

52 ARTICULATIONS Knee

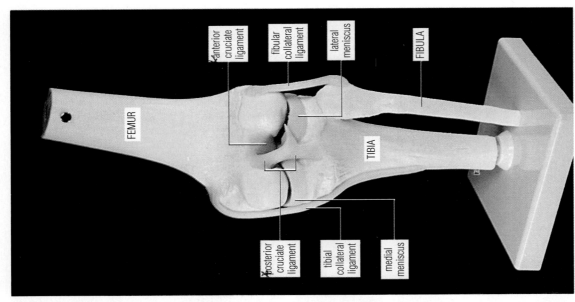

Figure 52c Posterior view of a model of the knee joint (0.50×). (Photo by D. Morton)

Figure 52b Lateral view of a model of the knee joint (0.50×). (Photo by D. Morton)

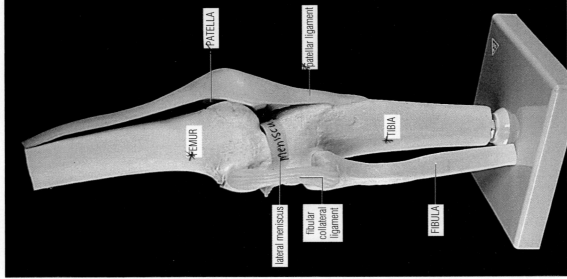

Figure 52a Anterior view of a model of the knee joint (0.50×). (Photo by D. Morton)

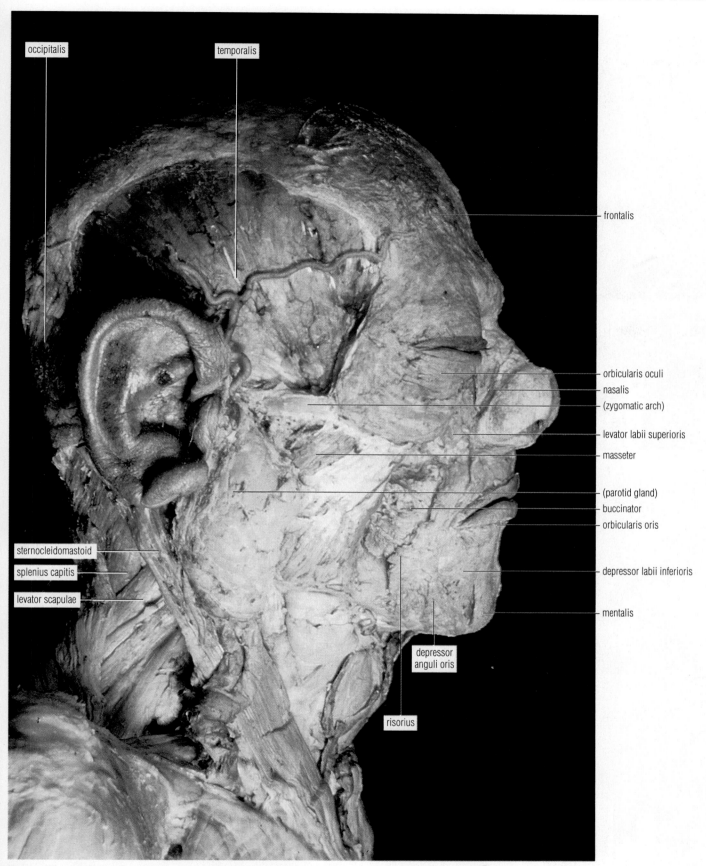

Figure 53 **Superficial muscles** of right **lateral head** and **neck.** Note that the zygomaticus minor and major and most of the platysma have been removed. (0.60×) (Reproduced by permission from *Human Anatomy and Physiology*, fifth edition, by J.W. Hole, Jr. Copyright Wm. C. Brown Publishers/McGraw-Hill Companies.)

54 Skeletal Muscles Neck

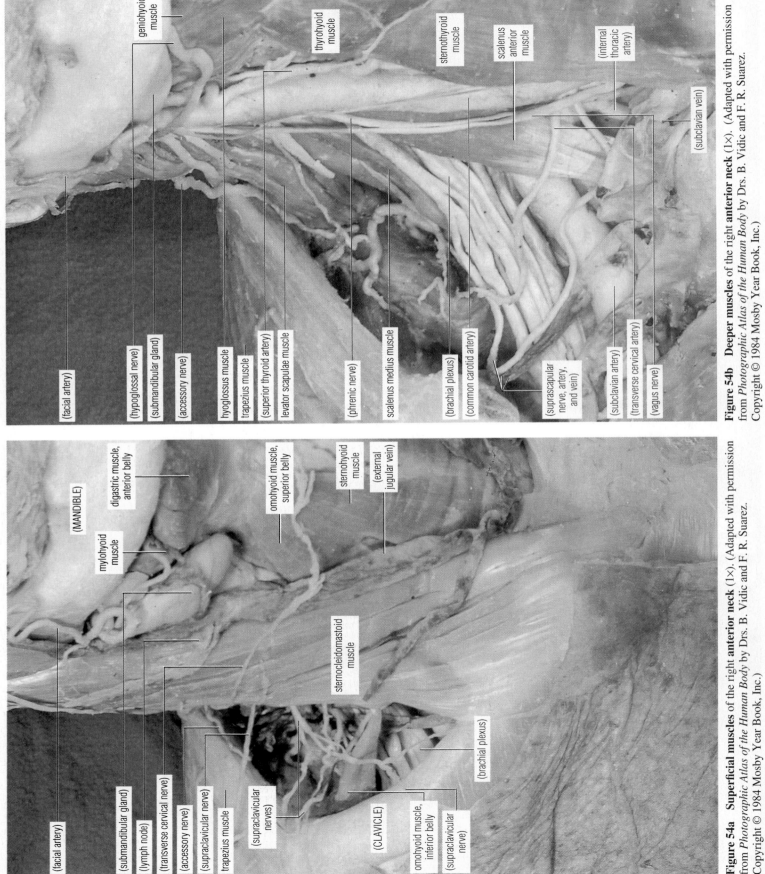

Figure 54b Deeper muscles of the right anterior neck (1×). (Adapted with permission from *Photographic Atlas of the Human Body* by Drs. B. Vidic and F. R. Suarez. Copyright © 1984 Mosby Year Book, Inc.)

Figure 54a Superficial muscles of the right anterior neck (1×). (Adapted with permission from *Photographic Atlas of the Human Body* by Drs. B. Vidic and F. R. Suarez. Copyright © 1984 Mosby Year Book, Inc.)

Upper Trunk and Arm — Skeletal Muscles

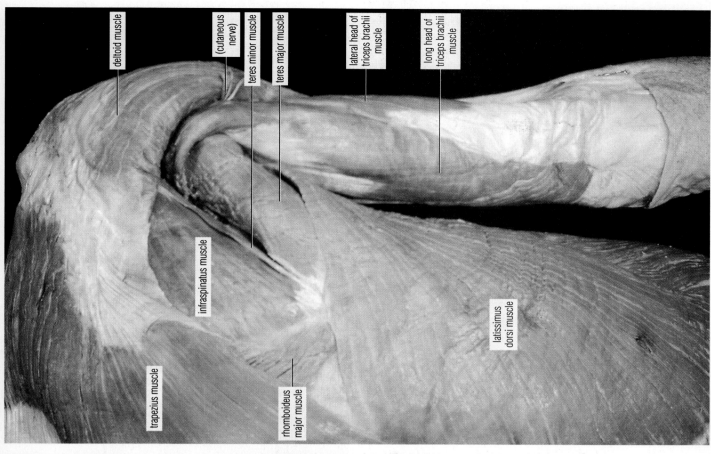

Figure 55b Superficial muscles of the right **posterior upper trunk** and **arm** (0.50×). (Adapted with permission from *Photographic Atlas of the Human Body* by Drs. B. Vidic and F. R. Suarez. Copyright © 1984 Mosby Year Book, Inc.)

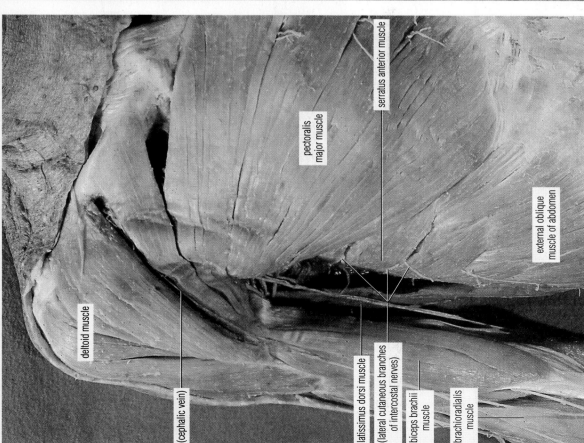

Figure 55a Superficial muscles of the right **anterior chest** and **arm** (0.50×). (Adapted with permission from *Photographic Atlas of the Human Body* by Drs. B. Vidic and F. R. Suarez. Copyright © 1984 Mosby Year Book, Inc.)

56 SKELETAL MUSCLES *Forearm and Hand*

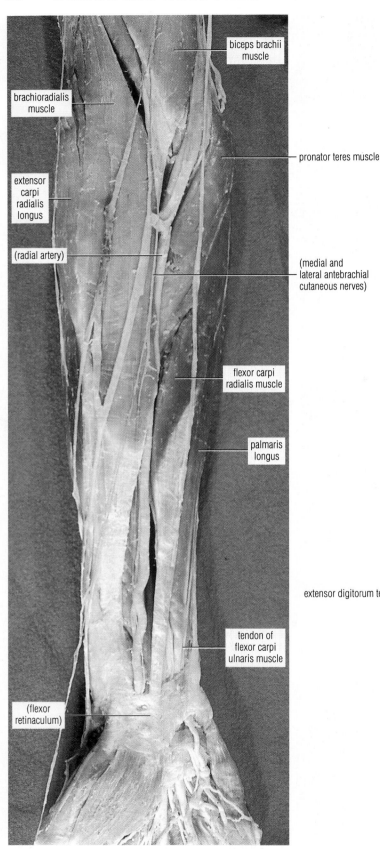

Figure 56a Superficial muscles of the right **anterior forearm** (0.50×). (Adapted with permission from *Photographic Atlas of the Human Body* by Drs. B. Vidic and F. R. Suarez. Copyright © 1984 Mosby Year Book, Inc.)

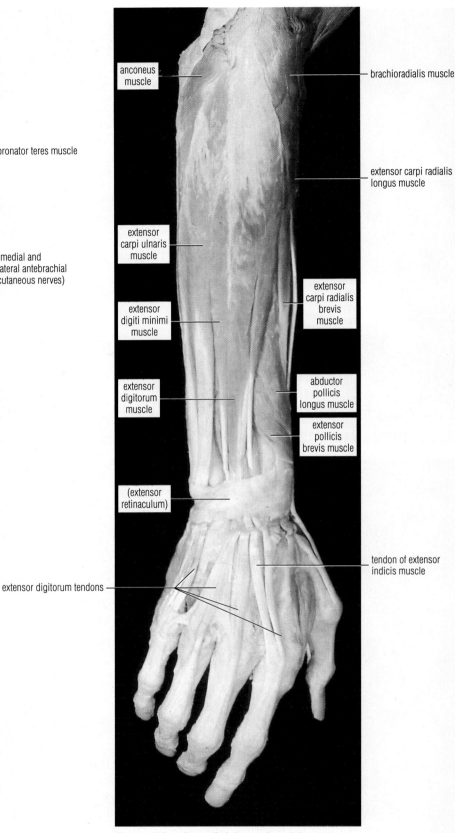

Figure 56b Superficial muscles of the right **posterior forearm** and **hand** (0.50×). (Adapted with permission from *Photographic Atlas of the Human Body* by Drs. B. Vidic and F. R. Suarez. Copyright © 1984 Mosby Year Book, Inc.)

Abdominal Region — Skeletal Muscles

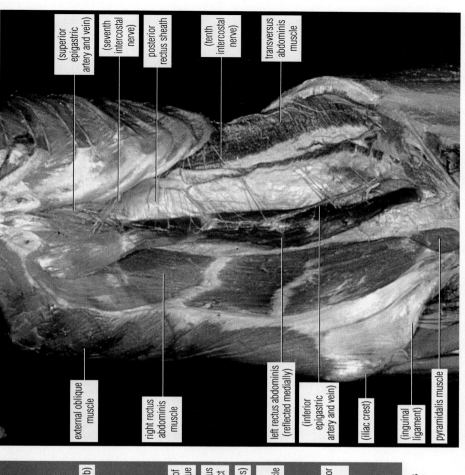

Figure 57b **Transversus abdominis.** This muscle is located under both the external and internal oblique muscles, which are removed from the left side (0.20×). (Courtesy of Dr. Robert Chase. Used by permission).

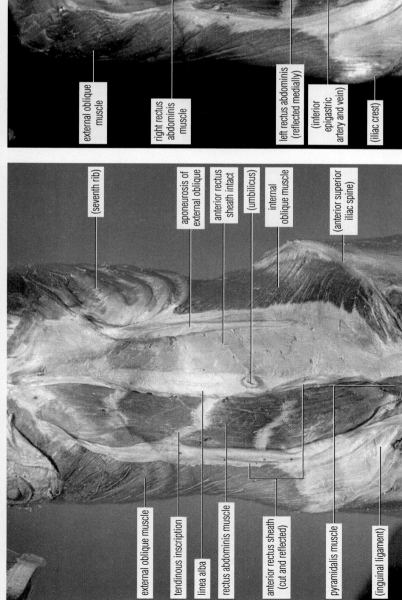

Figure 57a Abdominal region with the right **rectus abdominis** exposed by removal of its anterior sheath. Removal of the left external oblique has uncovered the left internal oblique (0.20×). (Courtesy of Dr. Robert Chase. Used by permission.)

58 SKELETAL MUSCLES *Gluteal Region*

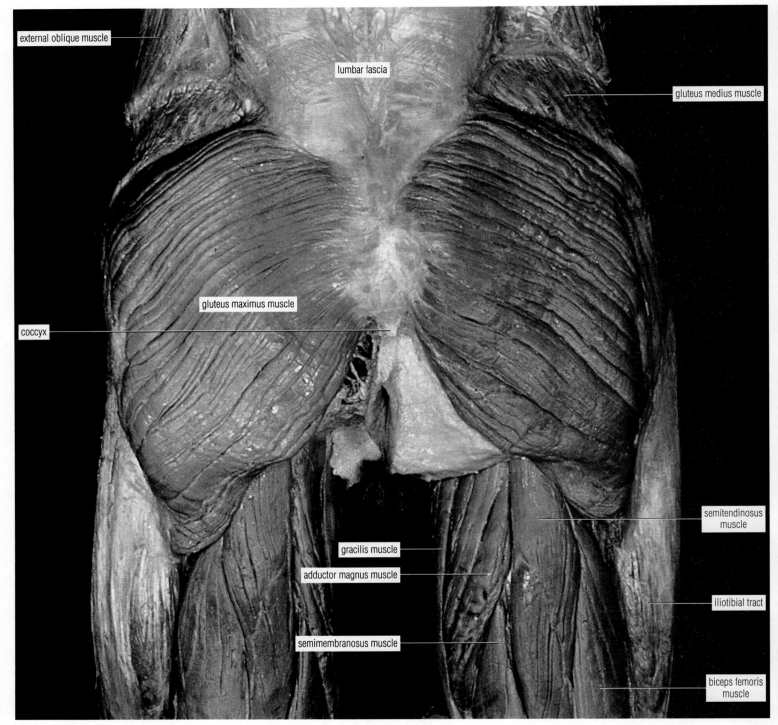

Figure 58 Gluteal muscles (0.35×). (Courtesy of Dr. Robert Chase. Used by permission.)

Thigh SKELETAL MUSCLES 59

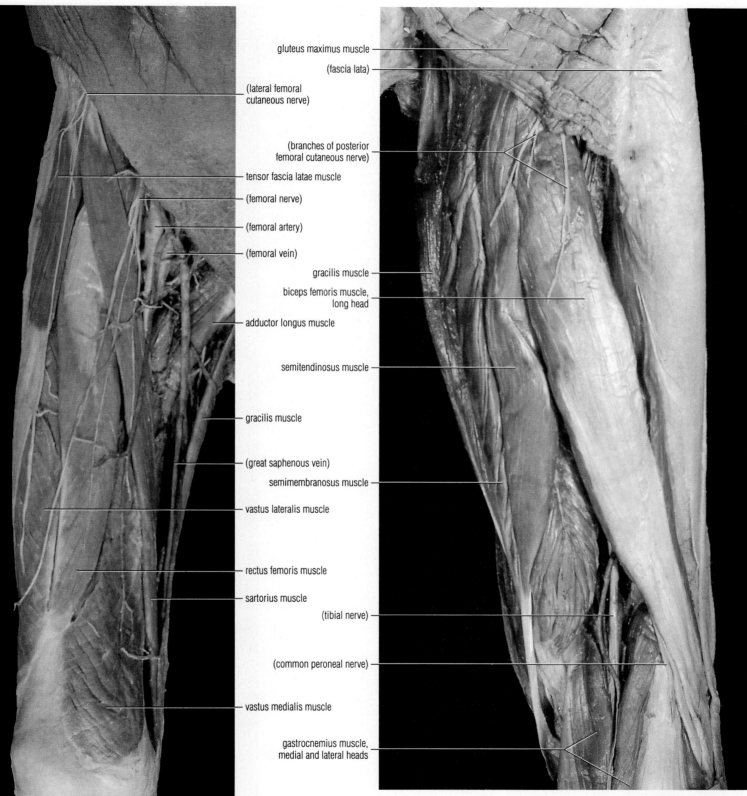

Figure 59a Superficial muscles of the right anterior and medial thigh (0.50×). (Adapted with permission from *Photographic Atlas of the Human Body* by Drs. B. Vidic and F. R. Suarez. Copyright © 1984 Mosby Year Book, Inc.)

Figure 59b Superficial muscles of the right posterior thigh (0.50×). (Adapted with permission from *Photographic Atlas of the Human Body* by Drs. B. Vidic and F. R. Suarez. Copyright © 1984 Mosby Year Book, Inc.)

60 SKELETAL MUSCLES   *Leg*

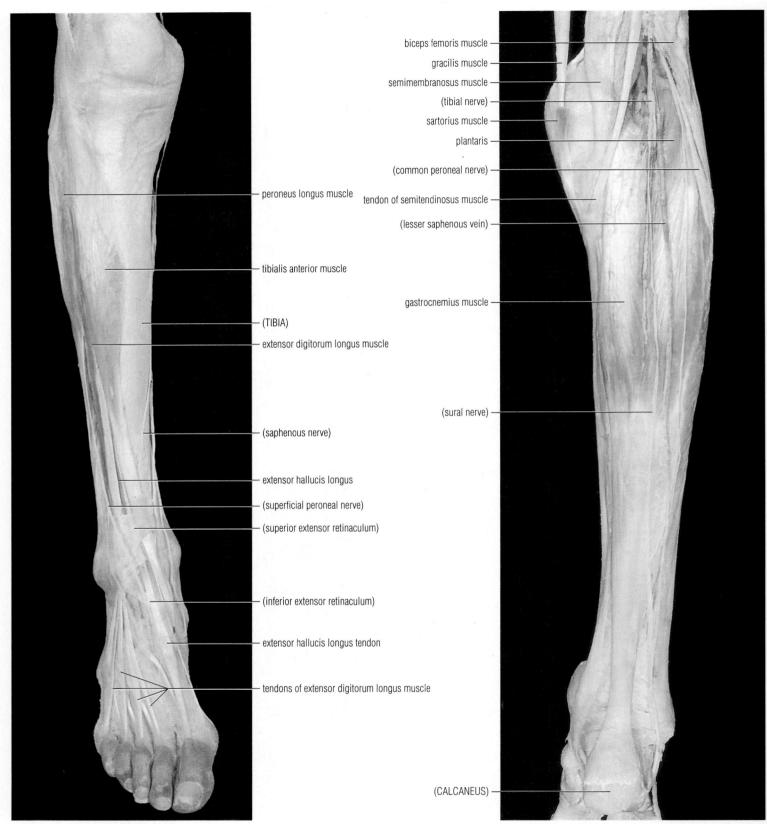

Figure 60a **Superficial muscles** of the right **anterior** and **medial leg** and the **dorsal** aspect of the **foot** (0.40×). (Adapted with permission from *Photographic Atlas of the Human Body* by Drs. B. Vidic and F. R. Suarez. Copyright © 1984 Mosby Year Book, Inc.)

Figure 60b **Superficial muscles** of the right **posterior leg** (0.40×). (Adapted with permission from *Photographic Atlas of the Human Body* by Drs. B. Vidic and F. R. Suarez. Copyright © 1984 Mosby Year Book, Inc.)

Human Brain Model — Nervous System

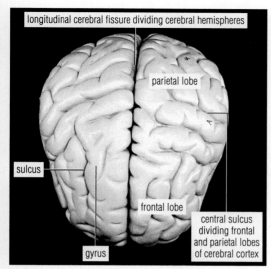

Figure 61a Superior view of a model of the human brain (0.40×). (Photo by D. Morton)

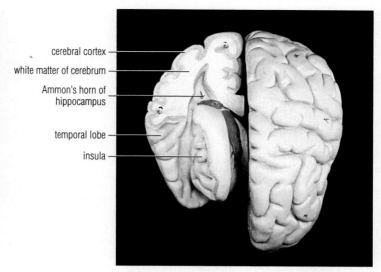

Figure 61b Superior view of a model of the human brain with a portion of its right side removed to show the fifth lobe of the cerebral cortex, the **insula** (0.40×). (Photo by D. Morton)

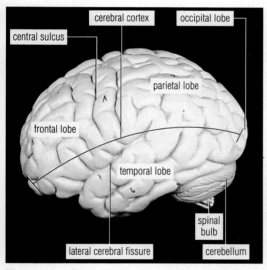

Figure 61c Lateral view of a model of the human brain (0.40×). (Photo by D. Morton)

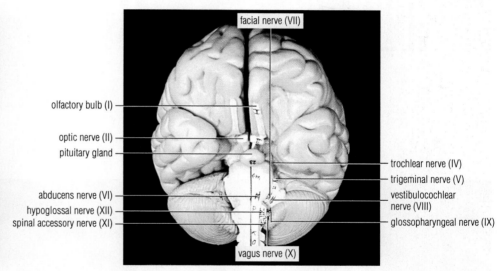

Figure 61d Inferior view of a model of the human brain with cranial nerves labeled (0.40×). (Photo by D. Morton)

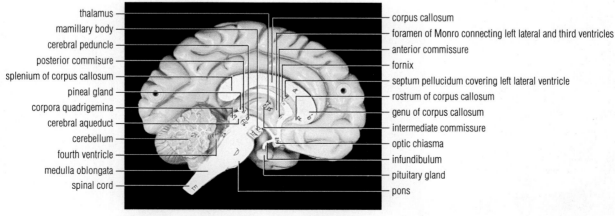

Figure 61e Midsagittal view of a model of the human brain (0.40×). (Photo by D. Morton)

Nervous System — Head/Spinal Cord

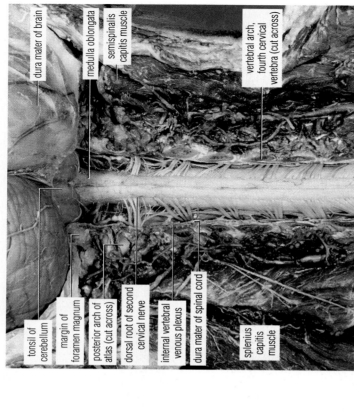

Figure 62b Origin of **spinal cord** (0.70×). (Photo courtesy of Dr. Robert Chase. Used by permission.)

Figure 62a Sagittal section of **head** (0.40×). (Reproduced by permission from *Human Anatomy and Physiology*, fifth edition, by J.W. Hole, Jr. Copyright Wm. C. Brown Publishers/McGraw-Hill Companies.)

Human Brain NERVOUS SYSTEM 63

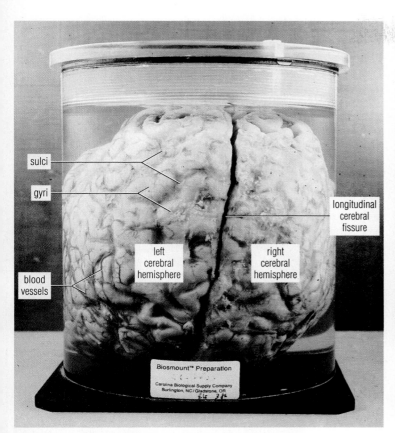

Figure 63a Superior view of the **human brain** (0.45×).
(Photo by D. Morton)

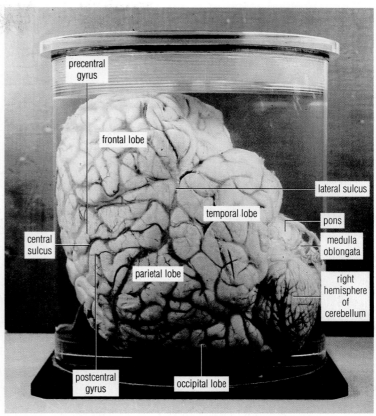

Figure 63b Lateral view of the **human brain** (0.40×).
(Photo by D. Morton)

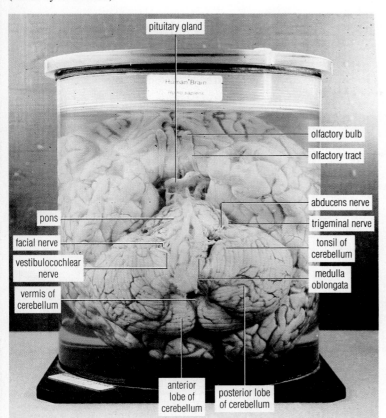

Figure 63c Inferior view of the **human brain** (0.40×).
(Photo by D. Morton)

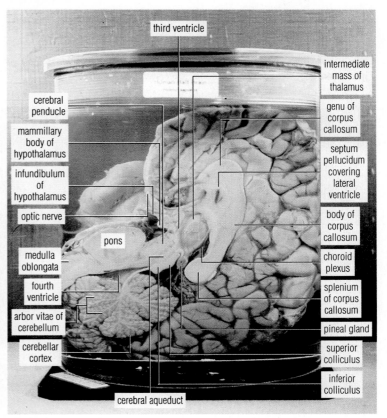

Figure 63d Sagittal section of the **human brain** (0.40×).
(Photo by D. Morton)

64 NERVOUS SYSTEM   *Sheep Brain*

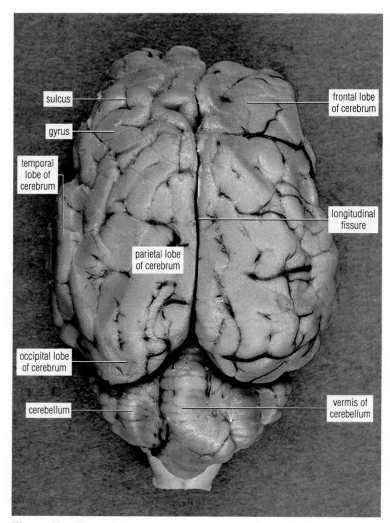

Figure 64a Dorsal view of **external anatomy** of **sheep brain** (1×). (Photo by D. Morton)

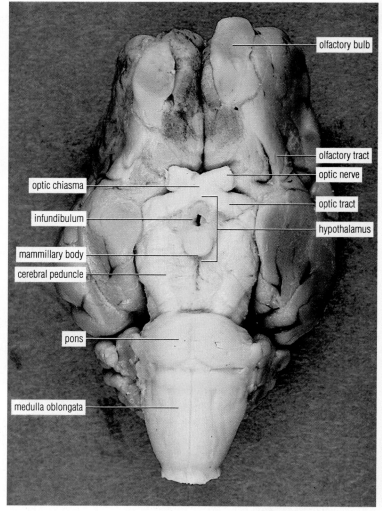

Figure 64b Ventral view of **external anatomy** of **sheep brain** (1×). (Photo by D. Morton)

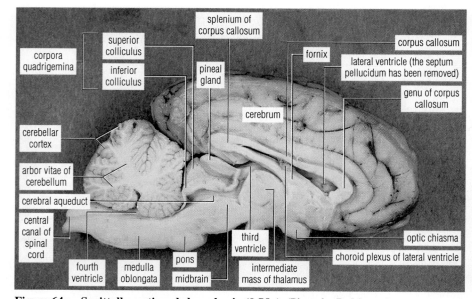

Figure 64c Sagittally sectioned sheep brain (0.75×). (Photo by D. Morton)

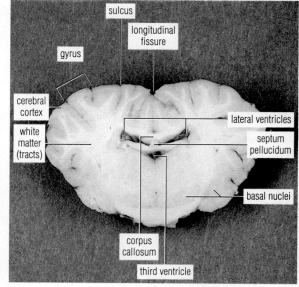

Figure 64d Anterior view of a **frontal** (coronal) **section** through the optic chiasma of **sheep brain** (0.90×). (Photo by D. Morton)

Eye/Human/Live and Models **RECEPTORS** 65

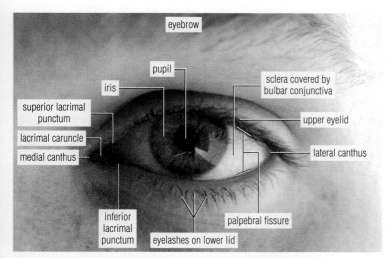

Figure 65a Living **eye** (1.7×). (Photo by D. Morton)

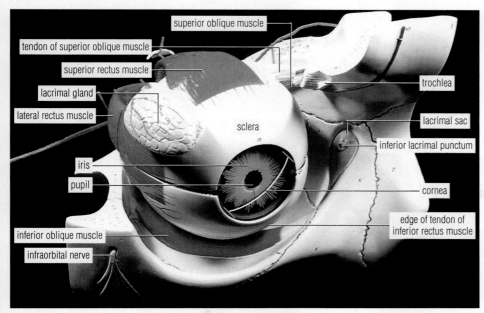

Figure 65b Model of eyeball in a portion of the orbit. The inferior rectus and medial rectus muscles cannot be seen (0.30×). (Photo by D. Morton)

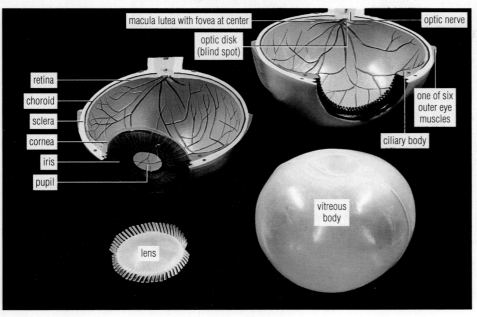

Figure 65c **Disassembled model** of an **eyeball** (0.30×). (Photo by D. Morton)

66 RECEPTORS *Eye/Sheep*

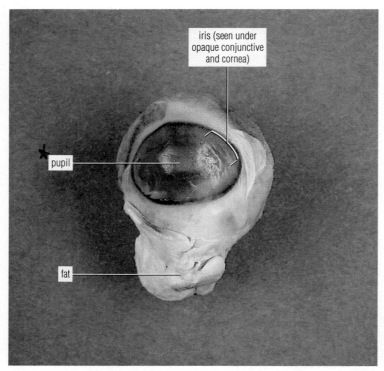

Figure 66a **External anatomy** of the **sheep eye** (0.95×). (Photo by D. Morton)

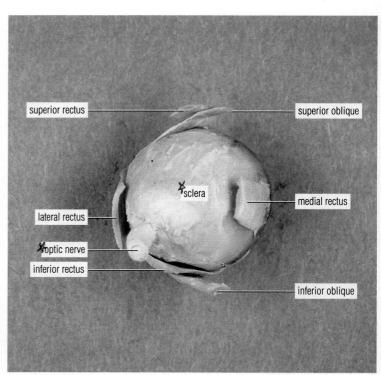

Figure 66b **External** (extrinsic) **eye muscles** of sheep (0.90×). (Photo by D. Morton)

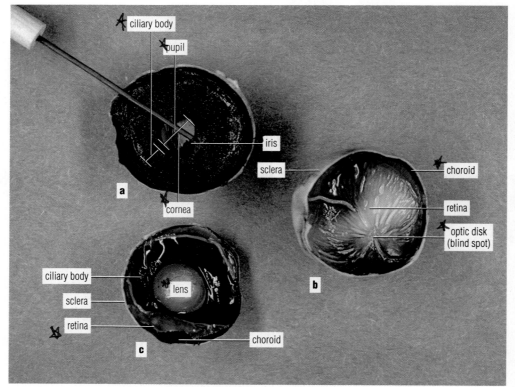

Figure 66c **Dissected sheep eye**: (a) inside of front of eyeball with lens removed; (b) inside of back of eyeball; (c) inside of front of eyeball (1.1×). (Photo by D. Morton)

Eye/Sections RECEPTORS

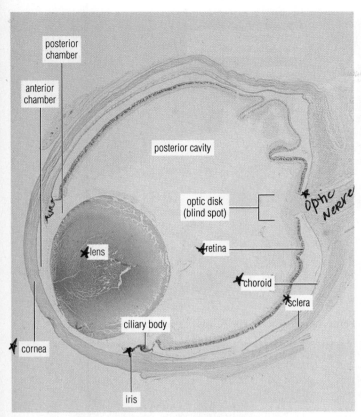

Figure 67a Sagittal section of **fetal eye** (prep. slide, 15×). (Photo by D. Morton) The anterior and posterior cavities comprise the anterior cavity.

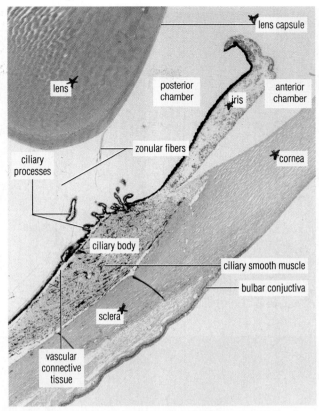

Figure 67b Sagittal section of the **ciliary body** and **iris** of a monkey eye (prep. slide, 30×). (Photo by D. Morton)

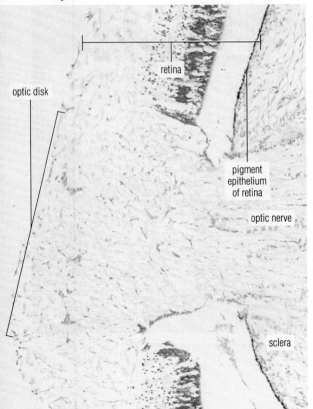

Figure 67c Sagittal section of the **optic nerve** and **optic disk** of a fetal eye (prep. slide, 100×). (Photo by D. Morton)

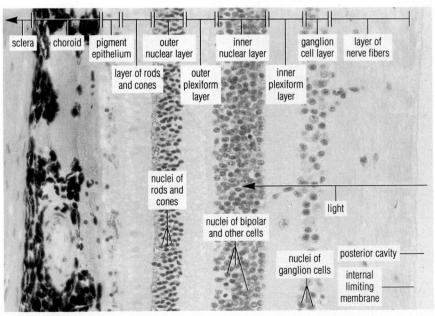

Figure 67d **Retina** of monkey eye (prep. slide, sec., 300×). (Photo by D. Morton)

68 RECEPTORS *Ear/Human/Live and Models*

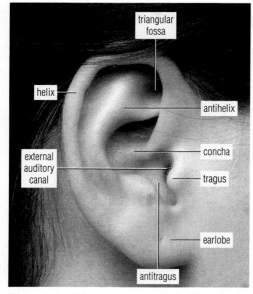

Figure 68a Living **outer ear** (1×). (Photo by D. Morton)

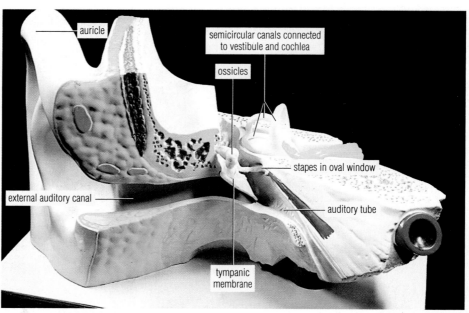

Figure 68c **Model** showing an **anterior view** of the ear (0.40×). (Photo by D. Morton)

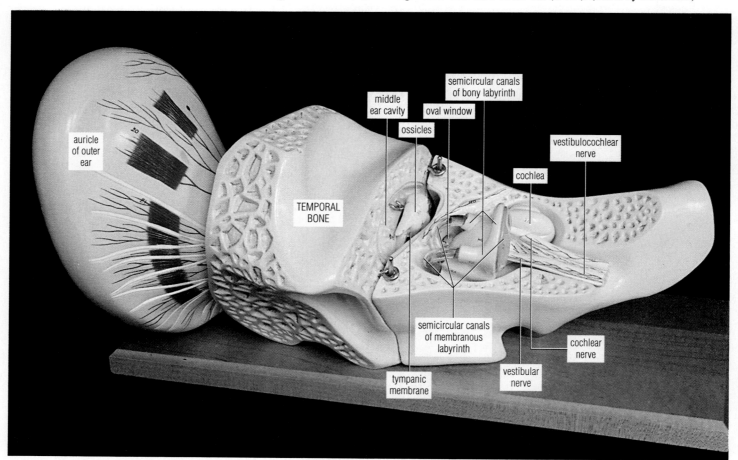

Figure 68b **Model** showing a **posterior view** of the **ear**. The middle ear and inner ear are embedded in the petrous portion of the temporal bone (0.60×). (Photo by D. Morton)

Cochlea/Taste Buds/Olfactory Epithelium — Receptors

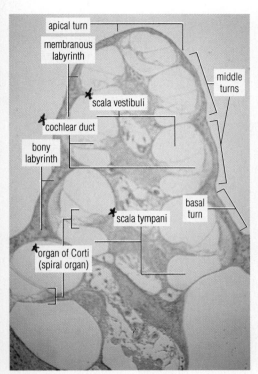

Figure 69a Section of the **cochlea** of the inner ear (prep. slide, 25×). (Photo by D. Morton)

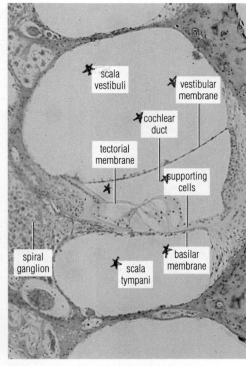

Figure 69b **Organ of Corti** (spiral organ) (prep. slide, c.s., 100×). (Photo by D. Morton)

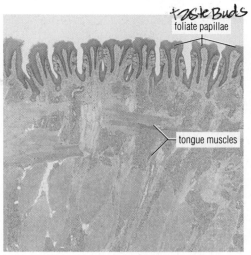

Figure 69c Foliate papillae in a section of **rabbit tongue** (prep. slide, 25×). (Photo by D. Morton)

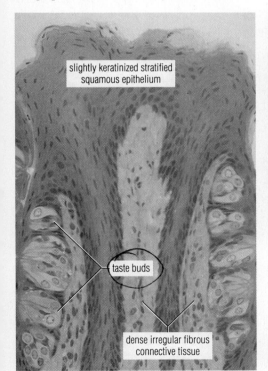

Figure 69d **Foliate papilla** with taste buds (prep. slide, l.s., 250×). (Photo by D. Morton)

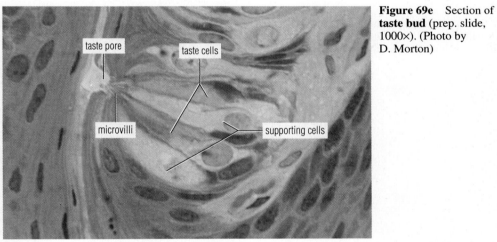

Figure 69e Section of **taste bud** (prep. slide, 1000×). (Photo by D. Morton)

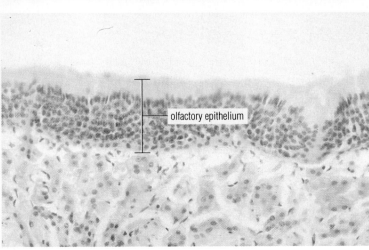

Figure 69f **Olfactory epithelium** (prep. slide, sec., 250×). (Photo by D. Morton)

ENDOCRINE SYSTEM — Pituitary Gland

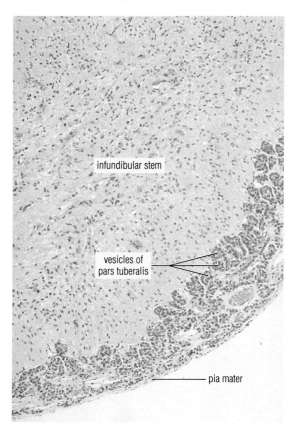

Figure 70a Infundibulum, which connects the hypothalamus to the pituitary gland (prep. slide, c.s., 100×). (Photo by D. Morton)

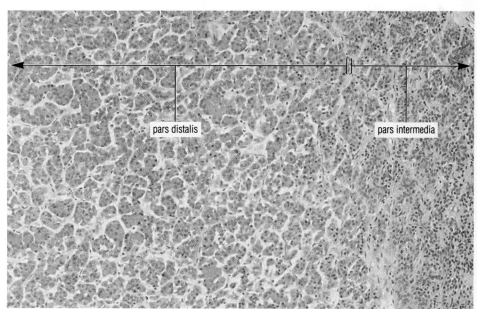

Figure 70b Anterior and **intermediate pituitary gland** of a cat (prep. slide, sec., 100×). (Photo by D. Morton)

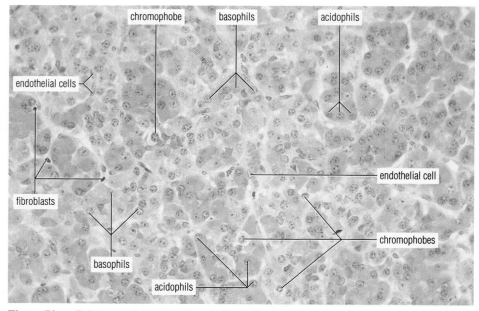

Figure 70c Cell types of the **anterior pituitary gland** of a cat (prep. slide, sec., 300×). (Photo by D. Morton)

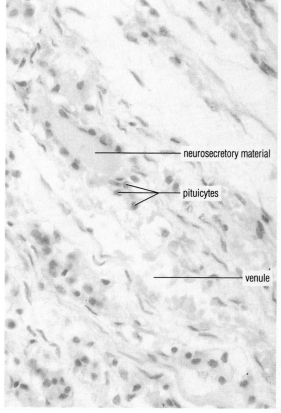

Figure 70d Posterior pituitary gland (prep. slide, sec., 250×). (Photo by D. Morton)

Pineal Gland/Thyroid Gland/Parathyroid Gland/Adrenal Gland — ENDOCRINE SYSTEM

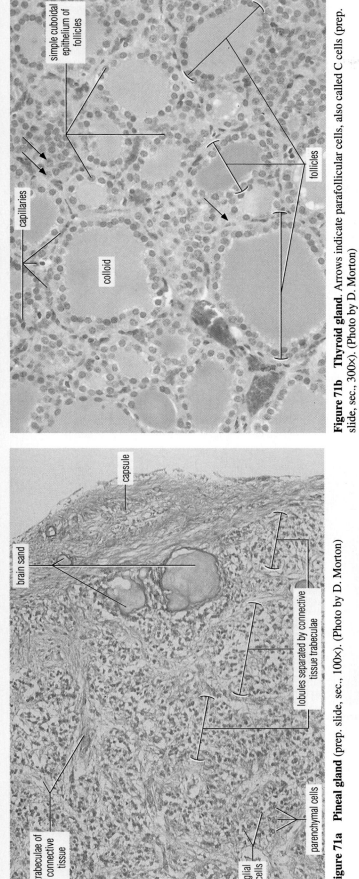

Figure 71a Pineal gland (prep. slide, sec., 100×). (Photo by D. Morton)

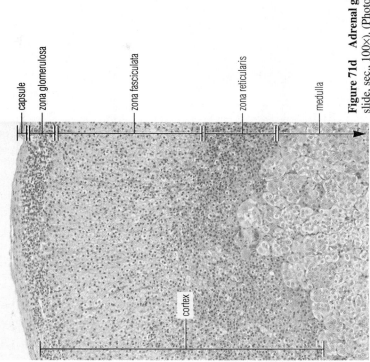

Figure 71b Thyroid gland. Arrows indicate parafollicular cells, also called C cells (prep. slide, sec., 300×). (Photo by D. Morton)

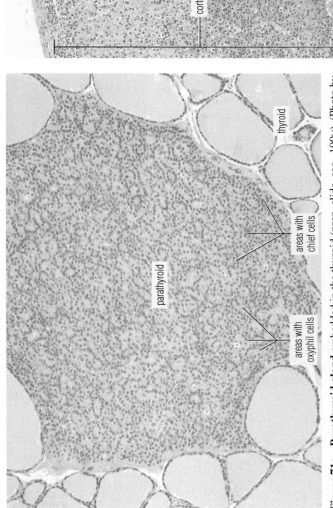

Figure 71c Parathyroid gland embedded in the thyroid (prep. slide, sec., 100×). (Photo by D. Morton)

Figure 71d Adrenal gland (prep. slide, sec., 100×). (Photo by D. Morton)

Digestive System — Tongue/Live and Papillae

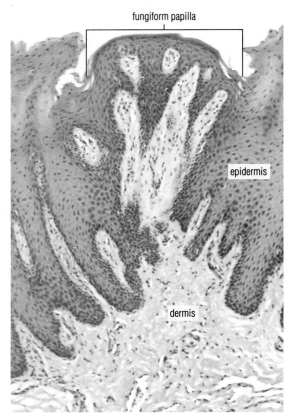

Figure 72a **Fungiform papilla** on the surface of **tongue** (prep. slide, sec., 250×). (Photo by D. Morton)

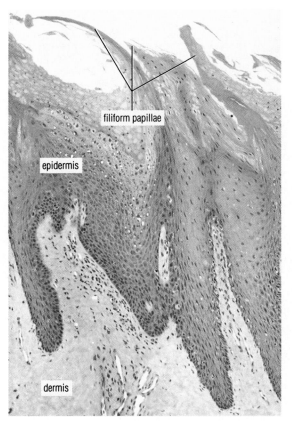

Figure 72b **Filiform papillae** on surface of **tongue** (prep. slide, sec., 250×). (Photo by D. Morton)

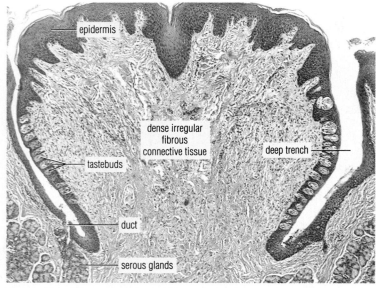

Figure 72c **Circumvallate papilla** on the surface of the **tongue** of a monkey (80×). (Photo courtesy of Dr. Michael H. Ross. Used by permission)

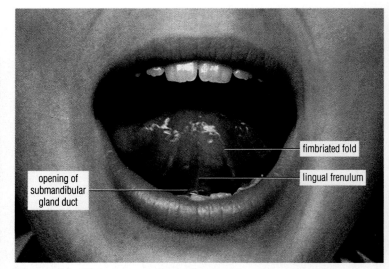

Figure 72d Living **tongue** and **oral cavity** (1×). (Photo by D. Morton)

Salivary Glands **DIGESTIVE SYSTEM** 73

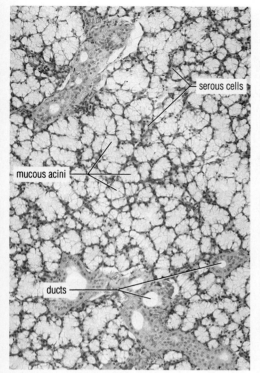

Figure 73a Sublingual gland. The secretion of this **salivary gland** is primarily mucous, and in humans serous acini are rare. Serous cells are incorporated into otherwise mucous acini (prep. slide, sec., 100×). (Photo by D. Morton)

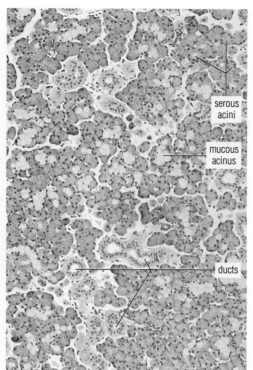

Figure 73b Submandibular gland. The secretion of this **salivary gland** is mixed, with mucous and serous acini present (prep. slide, sec., 100×). (Photo by D. Morton)

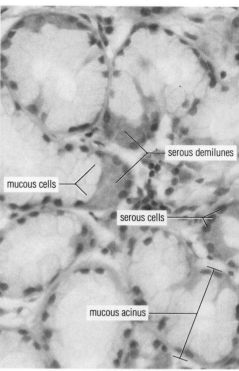

Figure 73c In both the sublingual and submandibular glands, **serous demilunes** cap the mucous acini (prep. slide, sec., 400×). (Photo by D. Morton)

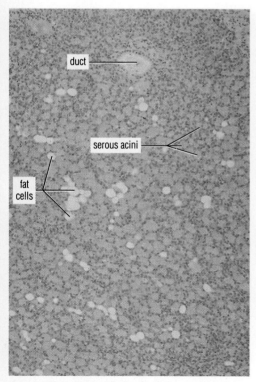

Figure 73d Parotid gland. Only serous acini are present (prep. slide, sec., 100×). (Photo by D. Morton)

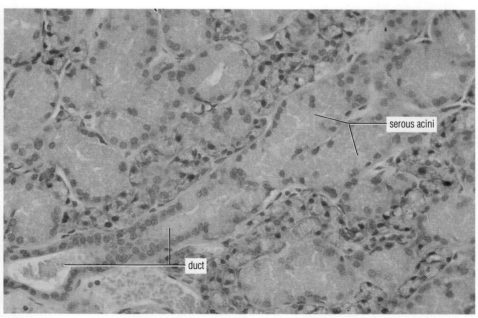

Figure 73e Longitudinally sectioned acini and duct in the **parotid gland** (prep. slide, sec., 300×). (Photo by D. Morton)

74 Digestive System — Tooth Development

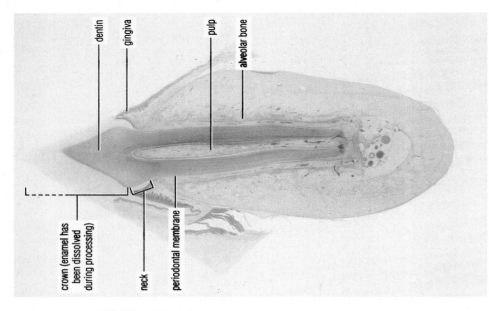

Figure 74c Mammalian **tooth** embedded in socket of alveolar ridge (l.s., 10×). (Photo by D. Morton)

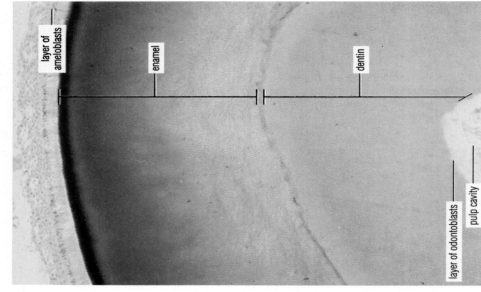

Figure 74b Late stage of **tooth development** (prep. slide, sec., 100×). (Photo by D. Morton)

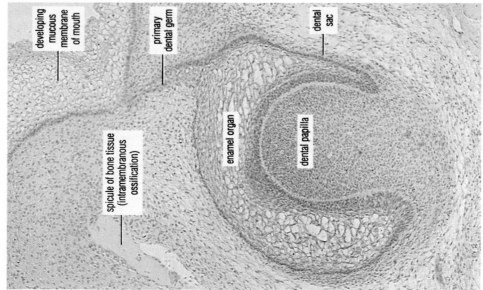

Figure 74a Early stage of **tooth development** (prep. slide, sec., 100×). (Photo by D. Morton)

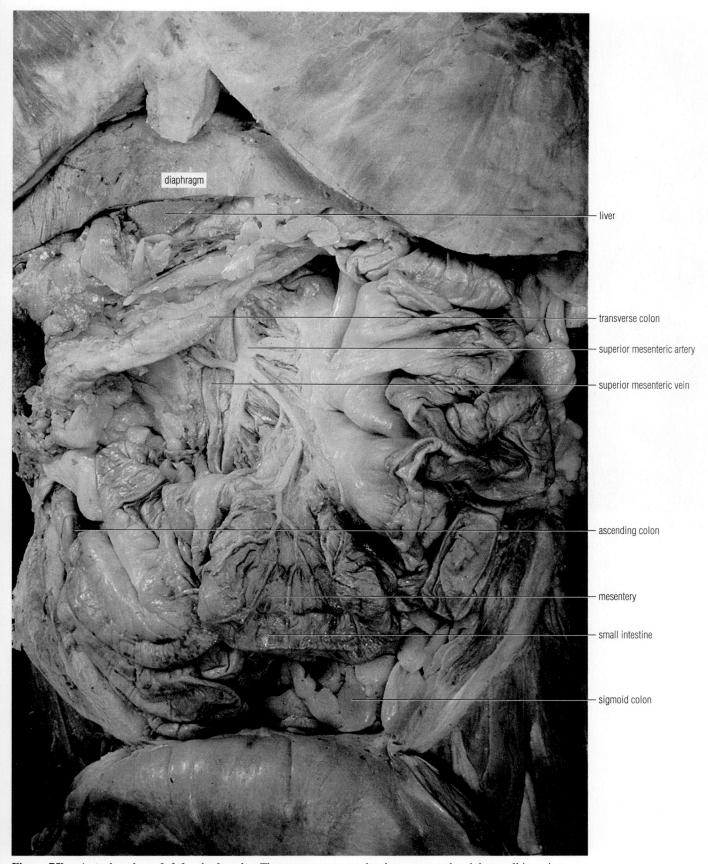

Figure 75b Anterior view of **abdominal cavity**. The greater omentum has been removed and the small intestine pushed to the left (0.40×). (Reproduced by permission from *Human Anatomy and Physiology*, fifth edition, by J.W. Hole, Jr. Copyright Wm. C. Brown Publishers/McGraw-Hill Companies.)

Digestive System — Esophagus, Stomach, and Duodenum

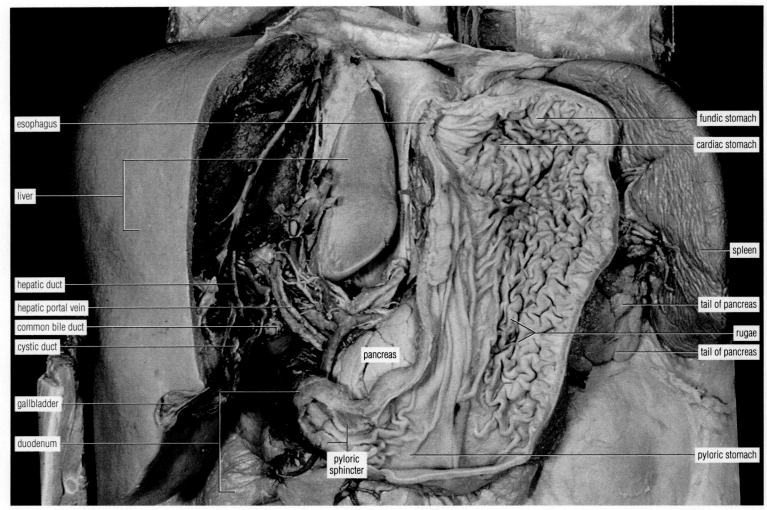

Figure 76 Opened **stomach**. (0.50×). (Photo courtesy of Dr. Robert Chase. Used by permission.)

Esophagus/Cardiac Stomach **DIGESTIVE SYSTEM**

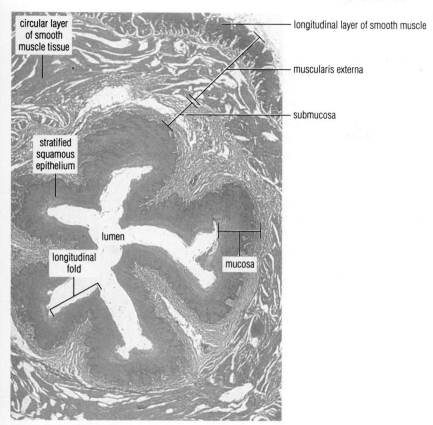

Figure 77a **Esophagus** (prep. slide, c.s., 10×). (Photo by D. Morton)

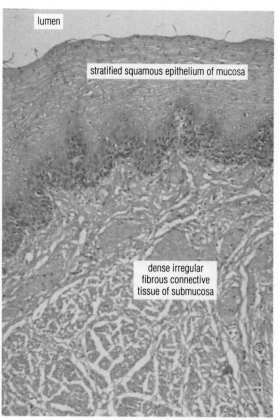

Figure 77b Stratified squamous **epithelium** of **esophagus** (prep. slide, sec., 100×). (Photo by D. Morton)

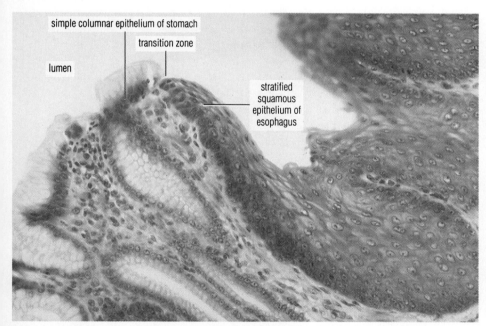

Figure 77c **Epithelial transition** between the **esophagus** and **stomach** (prep. slide, sec., 300×). (Photo by D. Morton)

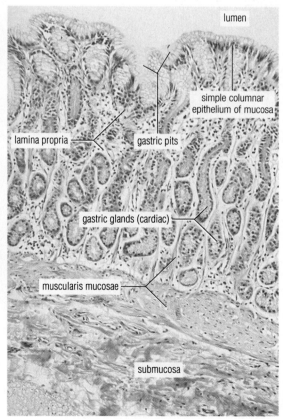

Figure 77d **Cardiac stomach** (prep. slide, sec., 100×). (Photo by D. Morton)

Digestive System — Fundic Stomach

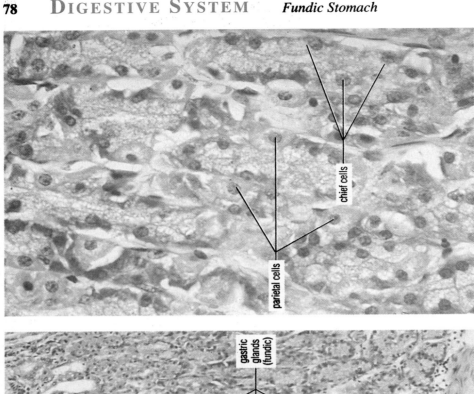

Figure 78c Fundic gastric glands contain **chief cells** and **parietal cells**. Chief cells secrete pepsinogen, which converts to pepsin, a proteolytic enzyme, when exposed to the HCL secreted by parietal cells. Parietal cells also secrete intrinsic factor, which is necessary for vitamin B_{12} absorption (prep. slide, sec., 500×). (Photo by D. Morton)

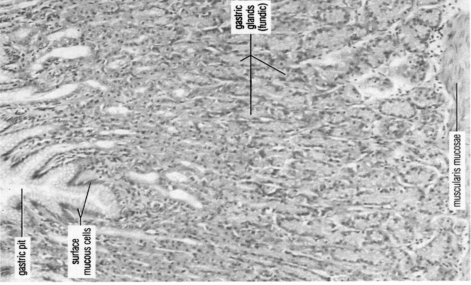

Figure 78b Mucosa of fundic stomach (prep. slide, sec., 100×). (Photo by D. Morton)

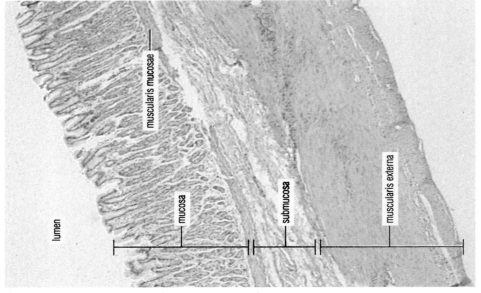

Figure 78a Fundic stomach (prep. slide, sec., 30×). (Photo by D. Morton)

Pyloric Stomach — DIGESTIVE SYSTEM

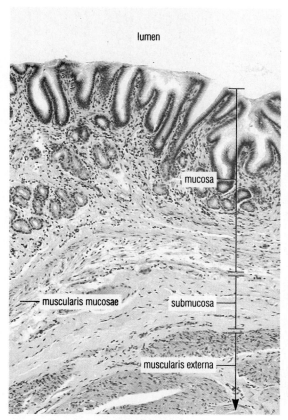

Figure 79a Pyloric stomach (prep. slide, sec., 100×). (Photo by D. Morton)

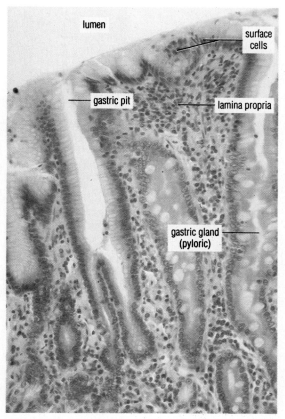

Figure 79b Mucosa of pyloric stomach (prep. slide, sec., 300×). (Photo by D. Morton)

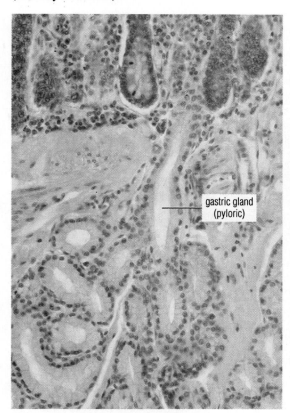

Figure 79c Pyloric gastric glands have a branched tubular structure (prep. slide, sec., 300×). (Photo by D. Morton)

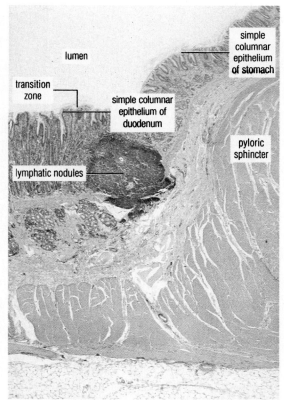

Figure 79d Pyloric sphincter (characterized by a thickened muscularis externa) and **epithelial transition** between the **stomach** and **duodenum** (prep. slide, sec., 30×). (Photo by D. Morton)

Digestive System — Small Intestine

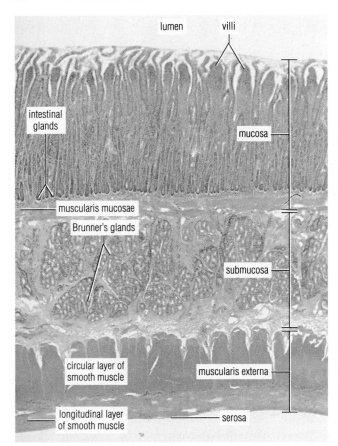

Figure 80a **Duodenum**, the initial portion of the **small intestine**, with characteristic **Brunner's glands** (l.s., 24×). (Photo by D. Morton)

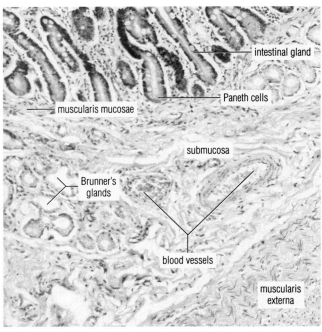

Figure 80b **Paneth cells** are present at the base of the intestinal glands. Among other substances, they secrete lysozyme, an antibacterial enzyme (prep. slide, sec., 100×). (Photo by D. Morton)

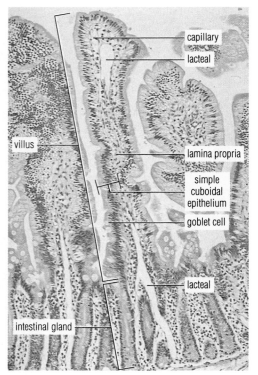

Figure 80c **Mucosa** of **jejunum**, the middle portion of the **small intestine**. Two portions of the same lacteal are shown (prep. slide, sec., 250×). (Photo by D. Morton)

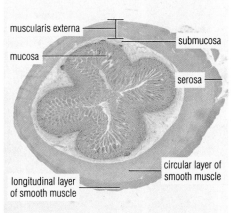

Figure 80d **Ileum**, the terminal portion of the **small intestine** (c.s., 6×). (Photo by D. Morton)

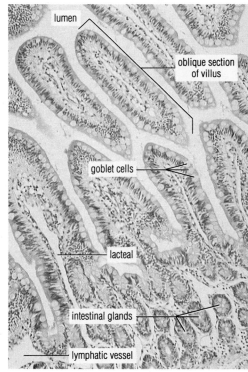

Figure 80e **Musosa** of **ileum** (prep. slide, sec., 250×). (Photo by D. Morton)

Large Intestine and Anal Canal — DIGESTIVE SYSTEM

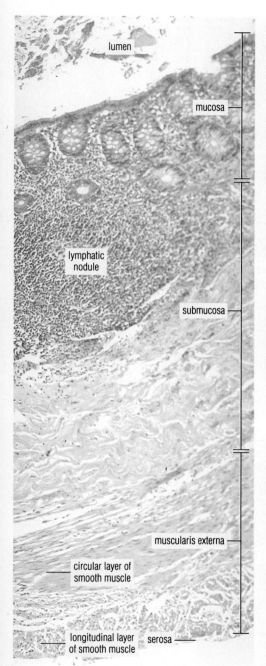

Figure 81a Colon of **large intestine** (c.s., 100×). (Photo by D. Morton)

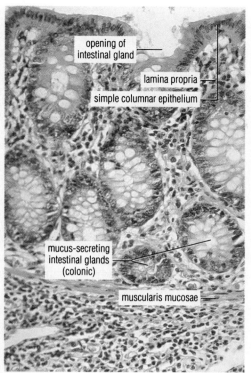

Figure 81b Mucosa of **colon** (prep. slide, sec., 250×). (Photo by D. Morton)

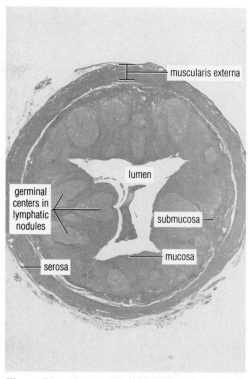

Figure 81c **Appendix**, a fingerlike diverticulum or pouch from the cecum, the initial portion of the **large intestine** (c.s., 10×). (Photo by D. Morton)

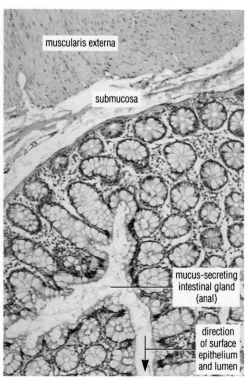

Figure 81d **Anal canal**, the terminal portion of the **large intestine** (prep. slide, sec., 250×). (Photo by D. Morton)

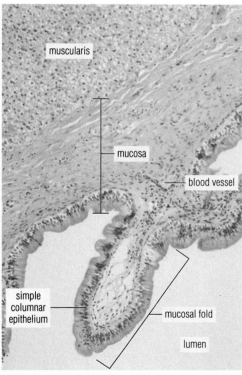

Figure 81e **Gallbladder** (prep. slide, sec., 100×). (Photo by D. Morton)

Digestive System — Liver and Pancreas

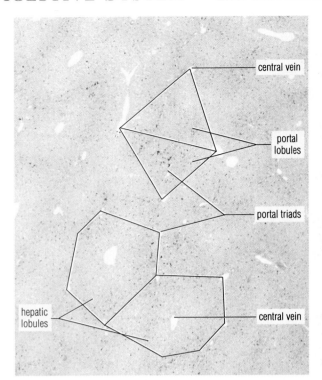

Figure 82a **Liver** from a mammal injected with a trypan blue suspension. Particles of this dye cannot enter living cells but are engulfed by Kupffer cells which are more numerous around the edges of the polygons. Each polygon outlines a **hepatic lobule** with **portal triads**—combinations of a portal vein, a branch of the hepatic artery, and a hepatic ductule or small bile duct—at several points around its edge and a central vein at its center. An alternative architecture emphasizing the exocrine functions of the liver is called a **portal lobule**. It places a portal triad at the center of a triangle whose sides connect the nearest three central veins. This section is stained with the dye nuclear fast red (prep. slide, sec., 20×). (Photo by D. Morton)

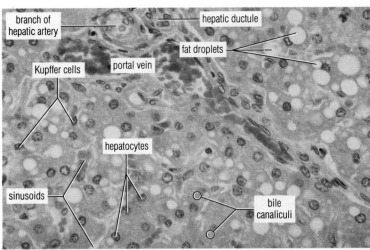

Figure 82c **Portal triad** and **hepatocytes**. **Bile canaliculi** are present between adjacent hepatocytes (prep. slide, sec., 400×). (Photo by D. Morton)

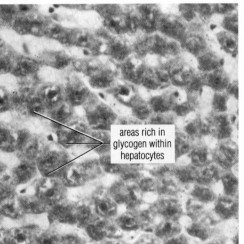

Figure 82d **Glycogen** (a storage form of glucose) is present in the hepatocytes of fed and rested individuals. This section is stained with PAS and hematoxylin (prep. slide, sec., 400×). (Photo by D. Morton)

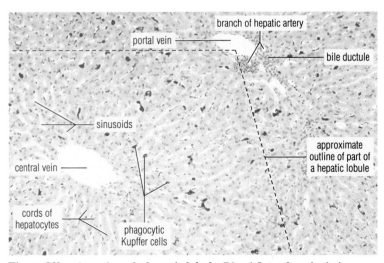

Figure 82b A portion of a **hepatic lobule**. Blood flows from both the **portal vein** and the **branch of the hepatic artery** through **sinusoids**, which drain into the **central vein**. Bile flows in the opposite direction to the blood in tiny bile canaliculi (see Figure 82c) toward the small **bile duct**. This section is stained with nuclear fast red (prep. slide, 100×). (Photo by D. Morton)

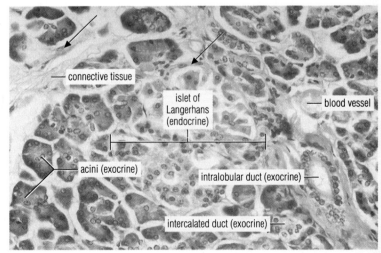

Figure 82e **Pancreas**. The arrows point out capillaries (prep. slide, sec., 250×). (Photo by D. Morton)

Human/Fetal Pig RESPIRATORY SYSTEM

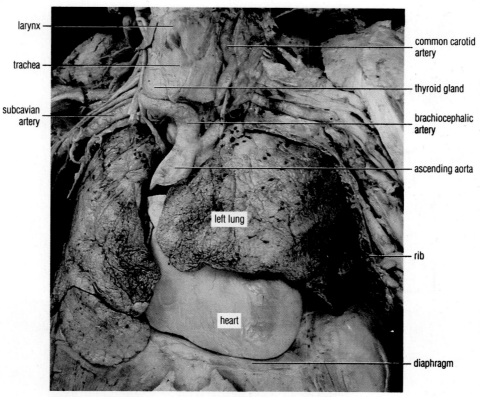

Figure 83a Anterior view of **thoracic cavity**. The brachiocephalic vein has been removed to show the aorta (0.20×). (Reproduced by permission from *Human Anatomy and Physiology*, fifth edition, by J.W. Hole, Jr. Copyright Wm. C. Brown Publishers/McGraw-Hill Companies.)

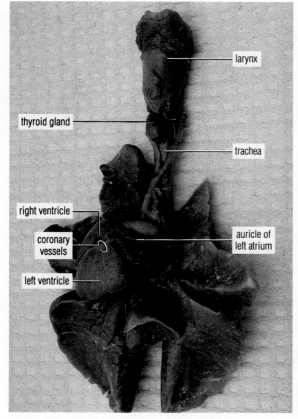

Figure 83b Ventral view of **respiratory system** of **fetal pig** (0.90×). (Photo by D. Morton)

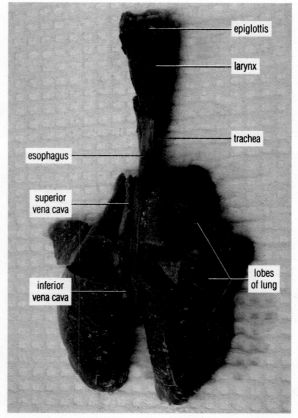

Figure 83c Dorsal view of **respiratory system** of **fetal pig** (0.90×). (Photo by D. Morton)

Respiratory System — Larynx/Models/Fetal Pig

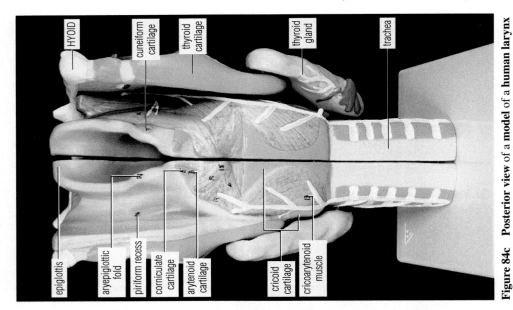

Figure 84c Posterior view of a model of a human larynx (0.60×). (Photo by D. Morton)

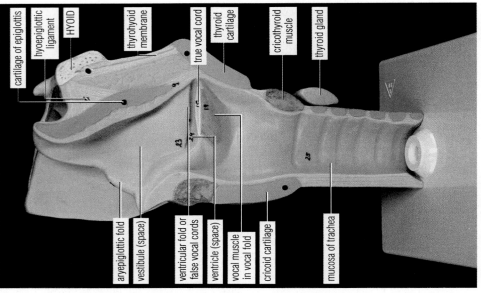

Figure 84b Medial view of a model of a human larynx (0.60×). (Photo by D. Morton)

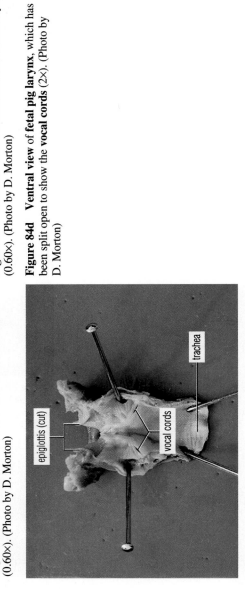

Figure 84d Ventral view of fetal pig larynx, which has been split open to show the vocal cords (2×). (Photo by D. Morton)

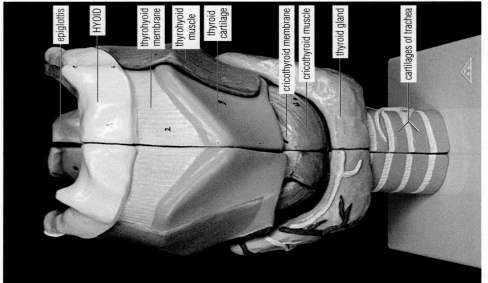

Figure 84a Anterior view of a model of a human larynx (0.60×). (Photo by D. Morton)

Trachea and Lung — RESPIRATORY SYSTEM

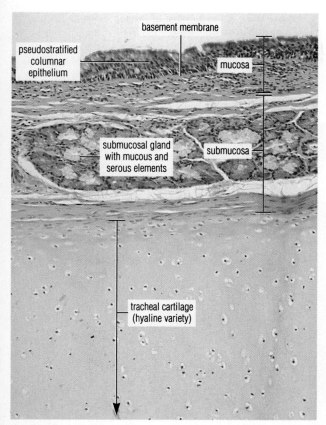

Figure 85a Wall of **trachea** (prep. slide, sec., 100×). (Photo by D. Morton)

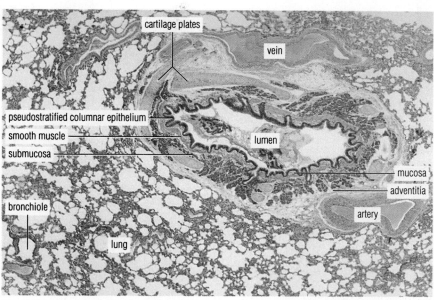

Figure 85b Secondary **bronchus** in **lung** (prep. slide, sec., 100×). (Photo by D. Morton)

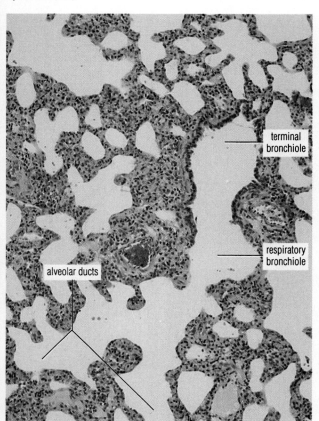

Figure 85c Terminal bronchiole, respiratory bronchiole, and **alveolar ducts** in **lung** (prep. slide, sec., 300×). (Photo by D. Morton)

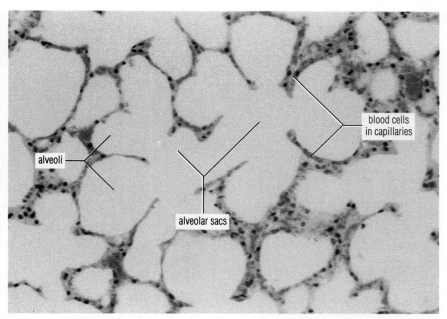

Figure 85d **Alveolar sacs** and **alveoli** in **lung** (prep. slide, sec., 500×). (Photo by D. Morton)

86 CIRCULATORY SYSTEM Heart

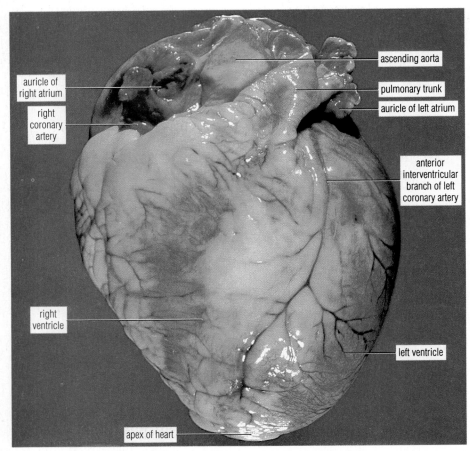

Figure 86a Anterior view of human heart (0.75×). (Photo © Martin Rotker/PHOTOTAKE.)

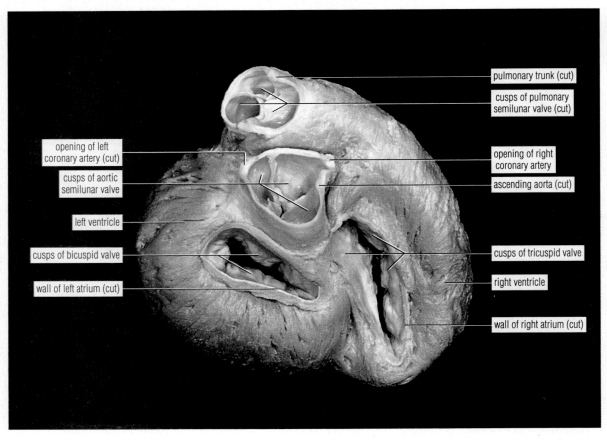

Figure 86b Superior view of base of human heart (0.75×). (Photo courtesy of Dr. Robert Chase. Used by permission.)

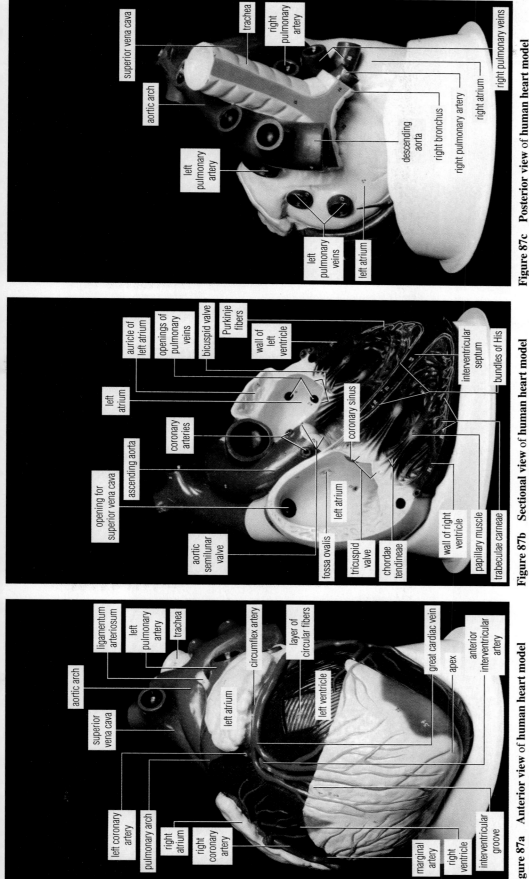

Figure 87a Anterior view of human heart model (0.40×). (Photo by D. Morton)

Figure 87b Sectional view of human heart model (0.35×). (Photo by D. Morton)

Figure 87c Posterior view of human heart model (0.35×). (Photo by D. Morton)

Circulatory System — Heart/Sheep

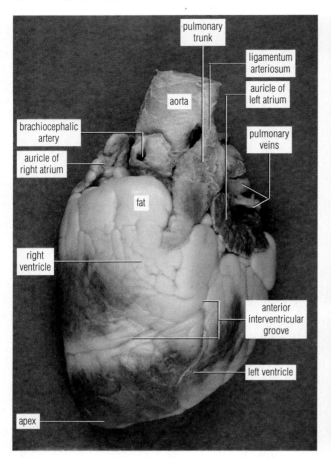
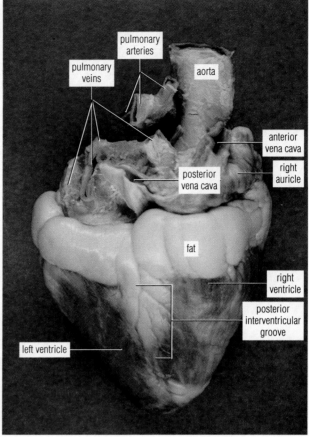

Figure 88a Ventral (left) and dorsal (right) surface views of sheep heart with dissected vessels (0.60×). (Photos by D. Morton)

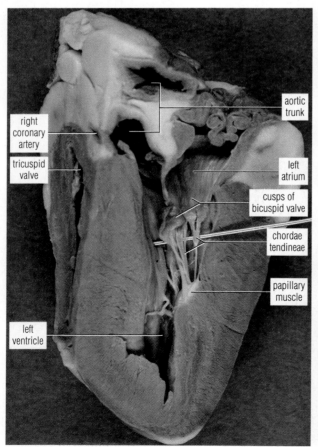
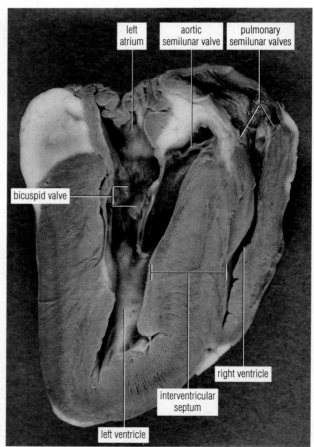

Figure 88b Ventral (left) and dorsal (right) views of sectioned sheep heart (0.60×). (Photos by D. Morton)

CIRCULATORY SYSTEM

Heart/Sections

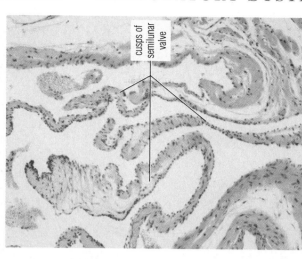

Figure 89c **Auricle**, a thin-walled flap forming part of each atrium of the **heart** (prep. slide, sec., 250×). (Photo by D. Morton)

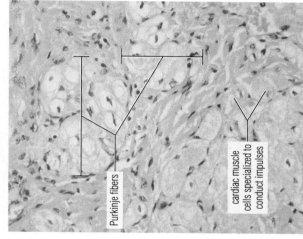

Figure 89e **Semilunar valve.** During diastole, semilunar valves prevent the backflow of blood from the aortic and pulmonary arches of the **heart** (prep. slide, sec., 100×). (Photo by D. Morton)

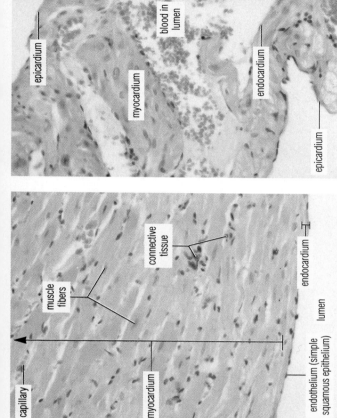

Figure 89b **Endocardium**, lining the inside of the wall of the **heart** (prep. slide, sec., 250×). (Photo by D. Morton)

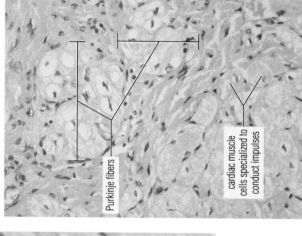

Figure 89d **Purkinje fibers**, part of an impulse-conducting system of specialized cardiac muscle cells that controls the contraction of the **heart** (prep. slide, sec., 250×). (Photo by D. Morton)

Figure 89a **Epicardium**, the outer layer of the **heart** wall, and **myocardium**, containing cardiac muscle fibers (cells) (prep. slide, sec., 400×). (Photo by D. Morton)

CIRCULATORY SYSTEM — Aorta

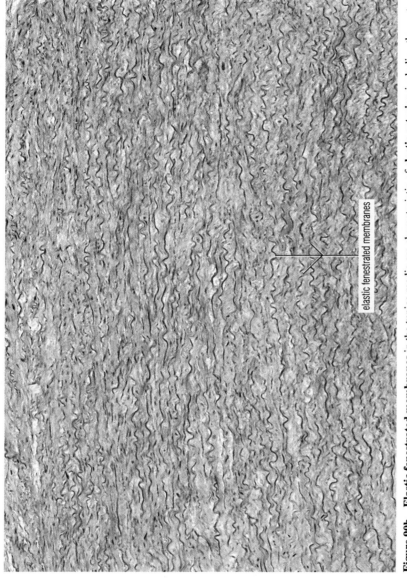

Figure 90b **Elastic fenestrated membranes** in the tunica media are characteristic of **elastic arteries**, including the aorta (prep. slide, c.s., 400×). (Photo by D. Morton)

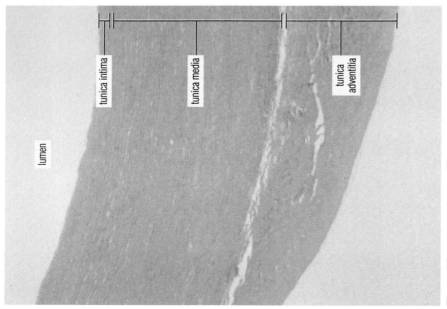

Figure 90a **Aorta.** (prep. slide, c.s., 3×). (Photo by D. Morton)

Vena Cava/Blood Vessels **CIRCULATORY SYSTEM**

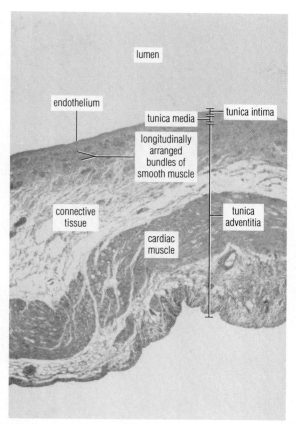

Figure 91a **Vena cava** (prep. slide, c.s., 12×). (Photo by D. Morton)

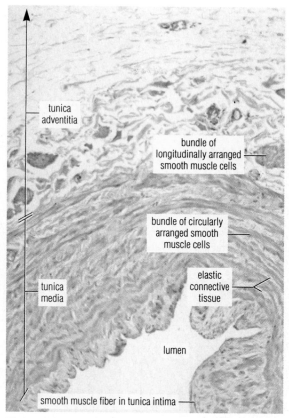

Figure 91b **Large vein** (prep. slide, c.s., 100×). (Photo by D. Morton)

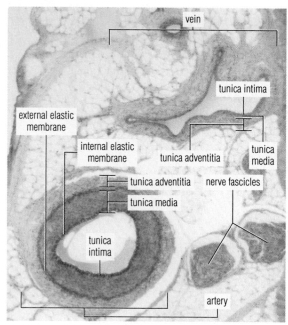

Figure 91c **Artery** and **vein** in a section stained for **elastic connective tissue**. The thickest tunica of arteries is the media. Veins have a relatively thicker tunica adventitia. Compared to their companion veins, arteries have thicker walls and narrower lumina, and their walls appear more organized (prep. slide, c.s., 35×). (Photo by D. Morton)

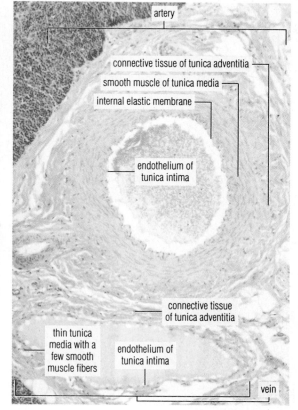

Figure 91d **Muscular artery** and companion **vein** (prep. slide, c.s., 100×). (Photo by D. Morton)

92 CIRCULATORY SYSTEM *Blood Vessels/Lymphatic Vessel*

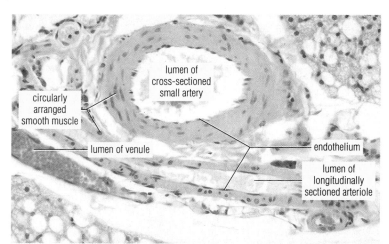

Figure 92a Small artery, **arteriole**, and **venule** (prep. slide, sec., 250×). (Photo by D. Morton)

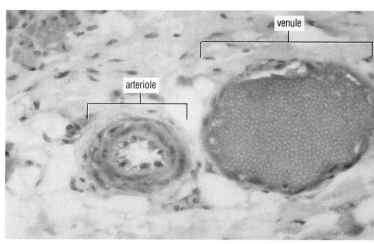

Figure 92b **Arteriole** and **venule** (prep. slide, c.s., 400×). (Photo by D. Morton)

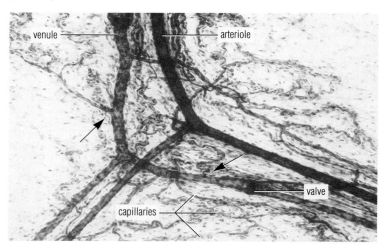

Figure 92c **Arteriole**, **venule**, and **capillary bed** in **mesentery**. Arrows indicate capillary joining venule (prep. slide, w.m., 100×). (Photo by D. Morton)

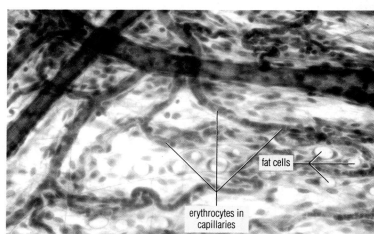

Figure 92d **Capillaries** (prep. slide, w.m., 400×). (Photo by D. Morton)

Figure 92e **Lymphatic vessel** with **valve** (prep. slide, w.m., 100×). (Photo by D. Morton)

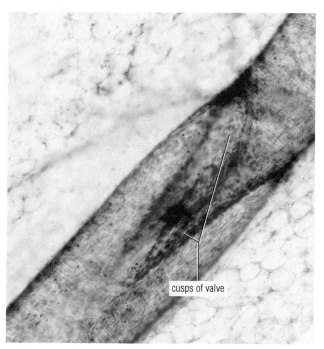

Blood # CIRCULATORY SYSTEM 93

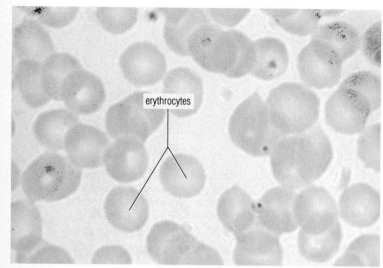

Figure 93a Smear of human blood: **erythrocytes** (prep. slide, w.m., 1600×). (Photo by D. Morton)

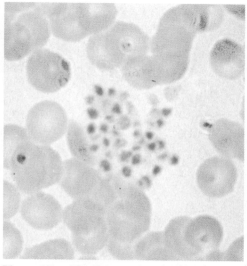

Figure 93b **Platelets** (prep. slide, w.m., 1500×). (Photo by D. Morton)

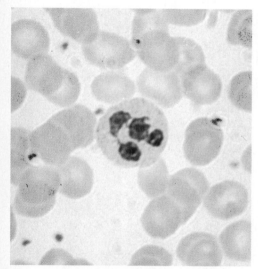

Figure 93c **Neutrophil** (prep. slide, w.m., 1800×). (Photo by D. Morton)

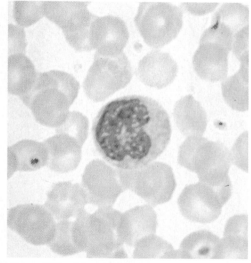

Figure 93d **Monocyte** (prep. slide, w.m., 1400×). (Photo by D. Morton)

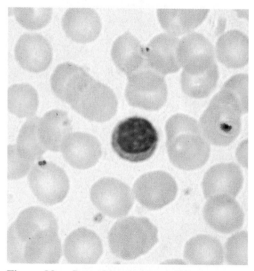

Figure 93e **Lymphocyte** (prep. slide, w.m., 2000×). (Photo by D. Morton)

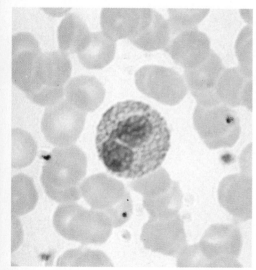

Figure 93f **Eosinophil** (prep. slide, w.m., 1300×). (Photo by D. Morton)

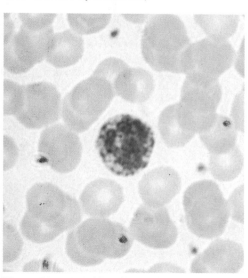

Figure 93g **Basophil** (prep. slide, w.m., 1100×). (Photo by D. Morton)

94 LYMPHATIC SYSTEM *Bone Marrow*

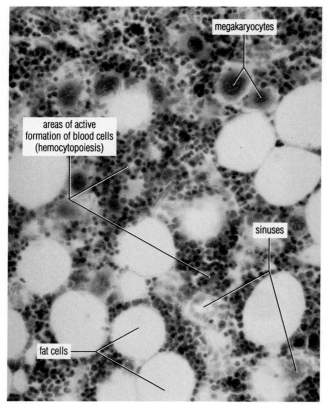

Figure 94a Bone marrow. Most blood cells form and develop in the red variety of bone marrow (prep. slide, sec., 100×). (Photo by D. Morton)

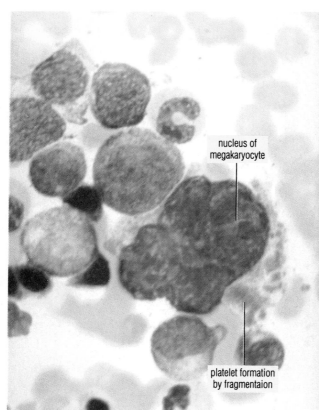

Figure 94b Megakaryocyte forming platelets (prep. slide, w.m., 1000×). (Photo by D. Morton)

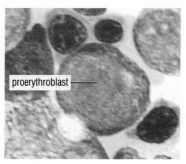

Figure 94c Proerythroblast, an early cell in the formation of erythrocytes (prep. slide, w.m., 1500×). (Photo by D. Morton)

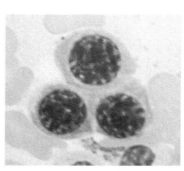

Figure 94e Polychromatic erythroblasts. Their cytoplasm is less blue because the hemoglobin they are synthesizing stains pink (prep. slide, w.m., 1500×). (Photo by D. Morton)

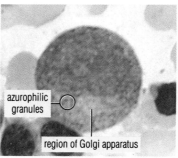

Figure 94g Promyelocyte, an early cell in the formation of granulocytes (prep. slide, w.m., 1500×). (Photo by D. Morton)

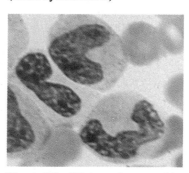

Figure 94i Metamyelocytes with indented nuclei (prep. slide, w.m., 1500×). (Photo by D. Morton)

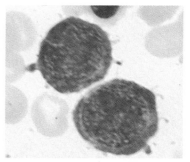

Figure 94d Basophilic erythroblasts with blue cytoplasm rich in ribosomes preparing for hemoglobin synthesis (prep. slide, w.m., 1500×). (Photo by D. Morton)

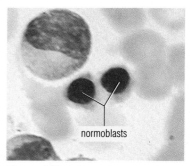

Figure 94f Normoblasts. These cells will extrude their nuclei to form young erythrocytes (prep. slide, w.m., 1500×). (Photo by D. Morton)

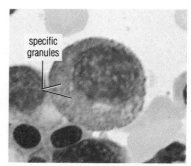

Figure 94h Myelocyte with a round nucleus and a few specific granules (prep. slide, w.m., 1500×). (Photo by D. Morton)

Thymus and Spleen LYMPHATIC SYSTEM

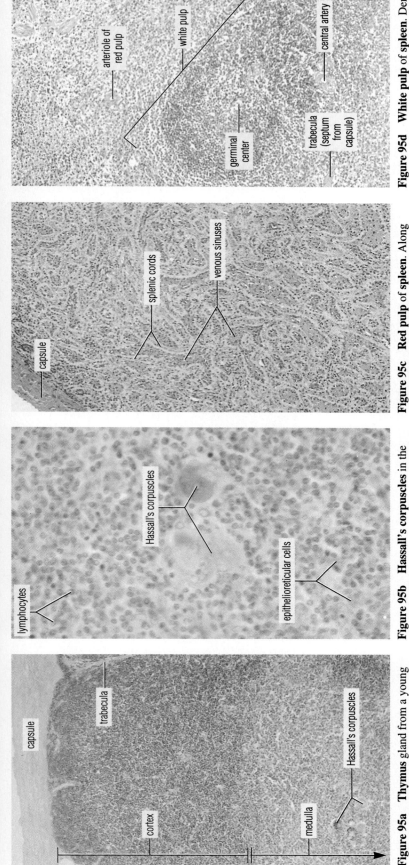

Figure 95a **Thymus** gland from a young individual. In general, lymphatic organs and structures contain populations of lymphocytes that exhibit various immunological functions. With puberty, the numbers of lymphocytes in the thymus, along with their supporting cellular reticulum, are considerably reduced (prep. slide, sec., 100×). (Photo by D. Morton)

Figure 95b **Hassall's corpuscles** in the medulla of the **thymus**. The thymus is the only lymphatic organ with a supporting mesh of epithelioreticular cells (prep. slide, sec., 400×). (Photo by D. Morton)

Figure 95c **Red pulp** of spleen. Along with immunological functions, the spleen filters the blood through a network of venous sinuses. Old and damaged erythrocytes are removed by macrophages in the splenic cords. These macrophages begin hemoglobin breakdown, including the release of iron, which is recycled. As is typical of most lymphatic organs, splenic structures are supported by a meshwork of reticular connective tissue (see Figure 14c) (prep. slide, sec., 100×). (Photo by D. Morton)

Figure 95d **White pulp** of spleen. Dense lymphatic tissue surrounds a central artery, which conveys blood to arterioles of the red pulp. **Germinal centers** within nodules are sites of active lymphocyte formation (prep. slide, sec., 100×). (Photo by D. Morton)

Lymphatic System — Tonsils, Appendix, and Lymph Node

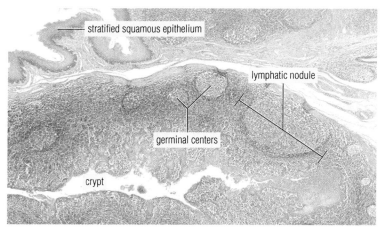

Figure 96a **Palatine tonsil**. Concentrations of nodular lymphatic tissue under the internal epithelia of the body are common, but there are a number of locations where they are invariably present. The palatine tonsils are located in the oral pharynx just behind the sides of the glossopalatine arch (prep. slide, sec., 25×). (Photo by D. Morton)

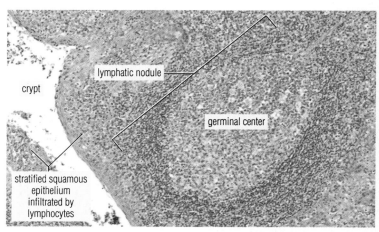

Figure 96b **Lymphatic nodule** from **palatine tonsil**. Typically, lymphocytes infiltrate the epithelium overlying dense lymphatic tissue (prep. slide, sec., 250×). (Photo by D. Morton)

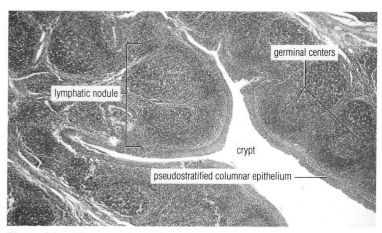

Figure 96c **Pharyngeal tonsil**, or **adenoid**, located on the posterior wall of the nasopharynx (prep. slide, sec., 20×). (Photo by D. Morton)

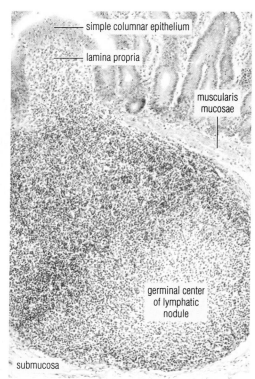

Figure 96d **Lymphatic nodule** from a **Peyer's patch** in the wall of the ileum (prep. slide, sec., 250×). (Photo by D. Morton)

Figure 96e **Lymphatic nodule** from the **appendix**. A low-power view of the appendix can be seen in Figure 79c (prep. slide, sec., 250×). (Photo by D. Morton)

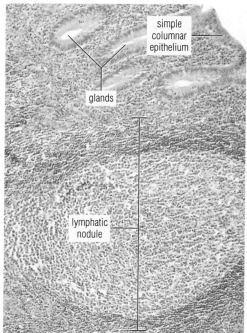

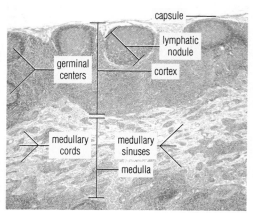

Figure 96f **Lymph node**. Smaller afferent lymphatic vessels join at lymph nodes to form larger efferent vessels. Lymph is filtered through the network of medullary sinuses, where macrophages in the medullary cords remove bacteria, malignant cells, and debris (prep. slide, sec., 20×). (Photo by D. Morton)

Kidney — URINARY SYSTEM

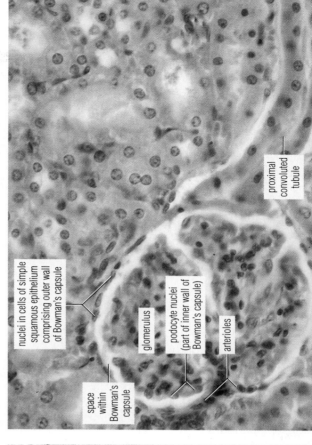

Figure 97b Midsagittal section of a nonhuman **mammalian kidney** with one renal papilla (prep. slide, 10×). (Photo by D. Morton)

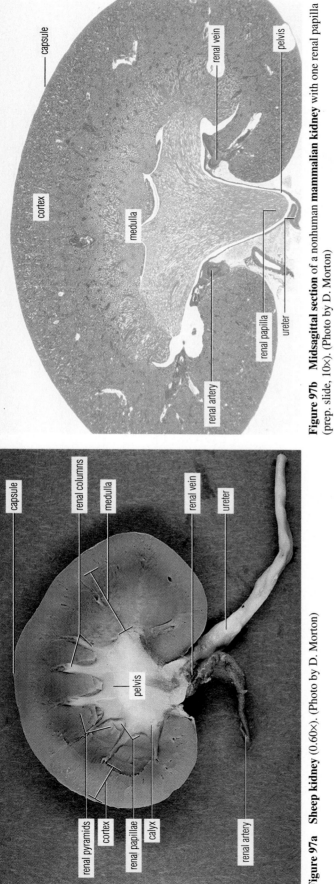

Figure 97a **Sheep kidney** (0.60×). (Photo by D. Morton)

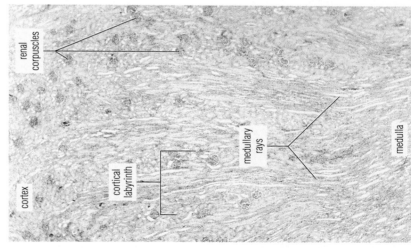

Figure 97d **Nephron.** A longitudinal section of the initial portion of a nephron shows a renal corpuscle, composed of **Bowman's capsule** and **glomerulus**, and a part of the **proximal convoluted tubule** (prep. slide, sec., 300×). (Photo by D. Morton)

Figure 97c **Cortex** and outer **medulla** of **kidney** (prep. slide, 25×). (Photo by D. Morton)

URINARY SYSTEM *Kidney*

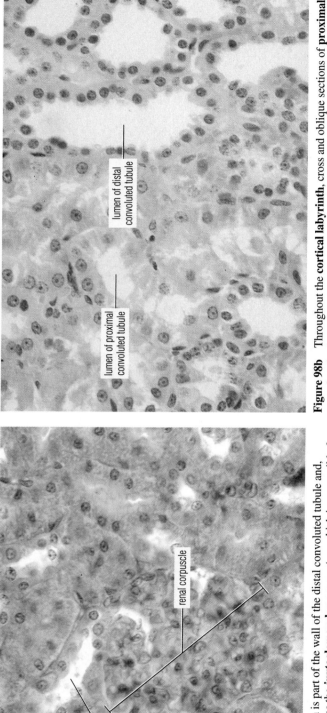

Figure 98a The **macula densa** is part of the wall of the distal convoluted tubule and, together with adjacent cells, forms the **juxtaglomerular apparatus**, which is responsible for the secretion of renin. Renin is an enzyme that initiates the renin-angiotensin negative feedback mechanism for the control of blood volume and blood pressure (prep. slide, sec., 250×). (Photo by D. Morton)

Figure 98b Throughout the **cortical labyrinth**, cross and oblique sections of **proximal convoluted tubules** and **distal convoluted tubules** predominate (prep. slide, sec., 250×). (Photo by D. Morton)

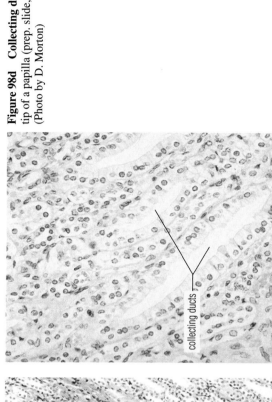

Figure 98d **Collecting ducts** near the tip of a papilla (prep. slide, sec., 250×). (Photo by D. Morton)

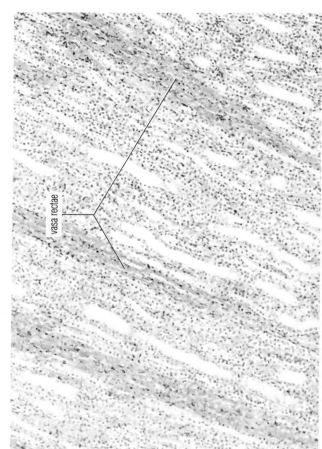

Figure 98c **Vasa rectae.** Arterioles from the efferent arterioles of some juxtamedullary renal corpuscles travel deep into the pyramids of the medulla, where they make hairpin turns, returning as venules. Groups of these vessels are called vena rectae (singular, vena recta) (prep. slide, sec., 100×). (Photo by D. Morton)

Ureter, Bladder, and Urethra URINARY SYSTEM

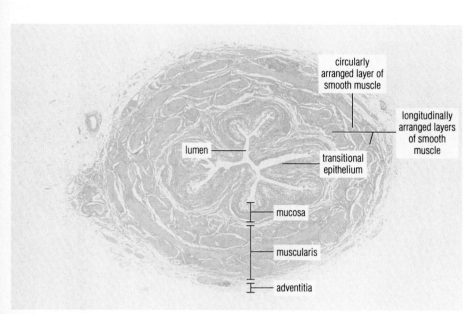

Figure 99a **Ureter** (prep. slide, c.s., 25×). (Photo by D. Morton)

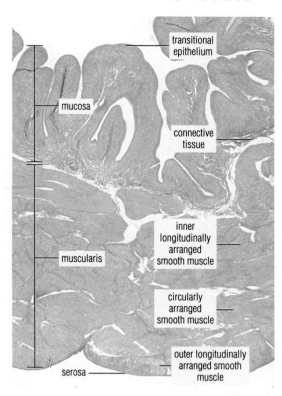

Figure 99b **Urinary bladder** (prep. slide, c.s., 20×). (Photo by D. Morton)

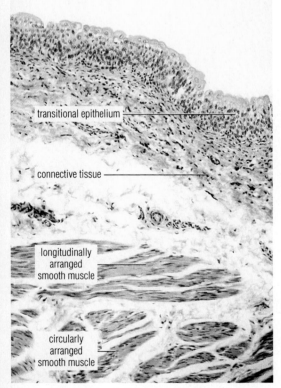

Figure 99c **Mucosa** of **urinary bladder** (prep. slide, l.s., 100×). (Photo by D. Morton)

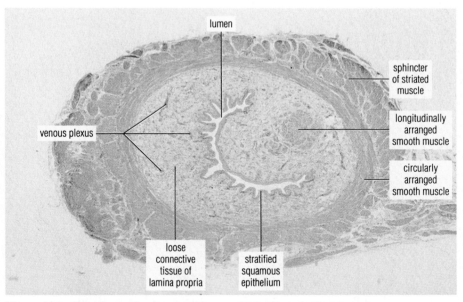

Figure 99d **Female urethra** (prep. slide, c.s., 50×). (Photo by D. Morton)

100 REPRODUCTIVE SYSTEM *Male/Model*

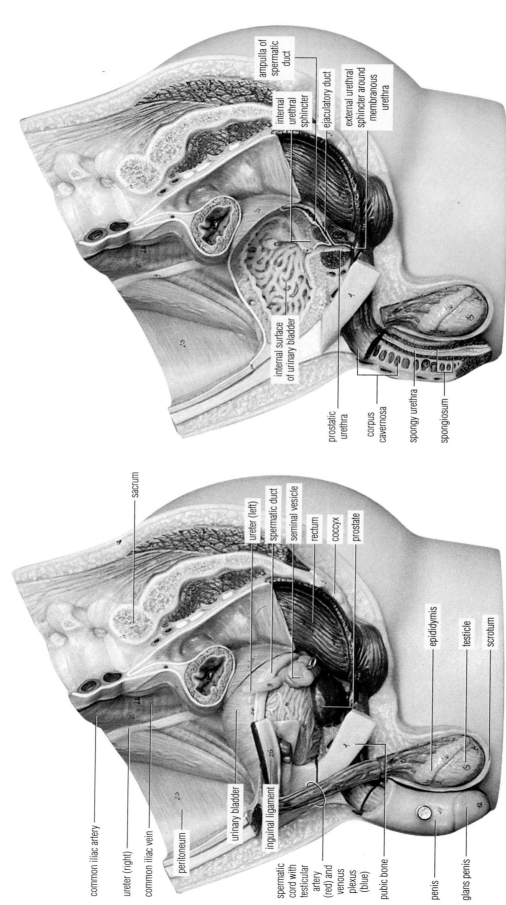

Figure 100a Model of **male reproductive system** and pelvis (0.40×). (Photo by D. Morton) **Figure 100b** Model of **male reproductive system** with left half of penis, urinary bladder, etc. removed (0.40×). (Photo by D. Morton)

Male/Testis — REPRODUCTIVE SYSTEM

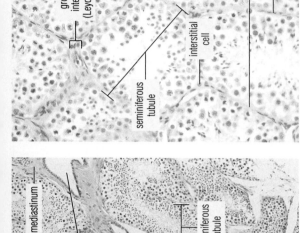

Figure 101c **Interstitial (Leydig) cells** situated between the seminiferous tubules secrete testosterone (prep. slide, sec., 200×). (Photo by D. Morton)

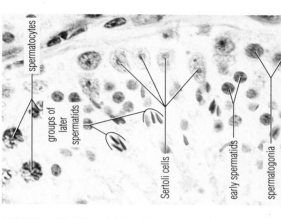

Figure 101e **Spermatogenesis.** Sertoli cells support spermatids as they develop into sperm (prep. slide, sec., 800×). (Photo by D. Morton)

Figure 101b The seminiferous tubules straighten out and drain sperm into the **rete testes**, an interconnected network of ducts within the mediastinum, a thickened region of the capsule (prep. slide, sec., 80×). (Photo by D. Morton)

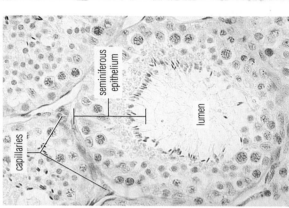

Figure 101d **Seminiferous epithelium** (prep. slide, sec., 300×). (Photo by D. Morton)

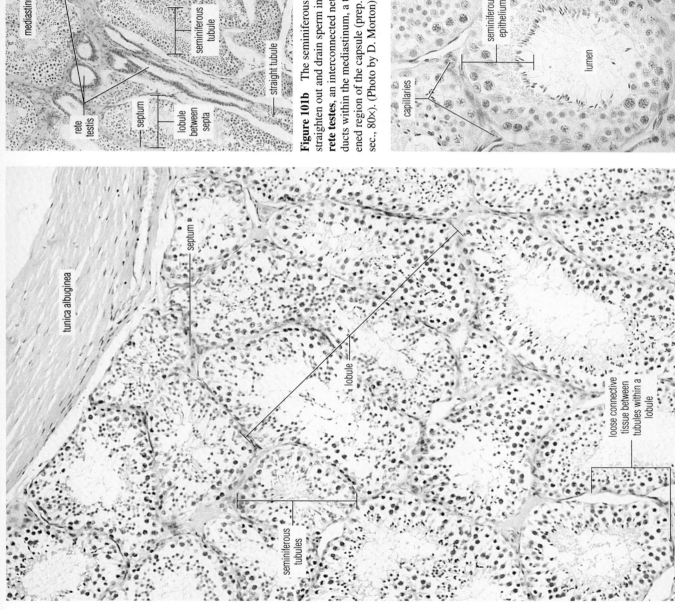

Figure 101a The **testis** is covered by the **tunica albuginea**, a thick capsule of fibrous connective tissue. Septa from the tunica albuginea divide the interior of the testis into several hundred lobules, each of which contains several highly folded **seminiferous tubules** (prep. slide, sec., 150×). (Photo by D. Morton)

REPRODUCTIVE SYSTEM — Male/Epididymus, Sperm, and Seminal Vesicle

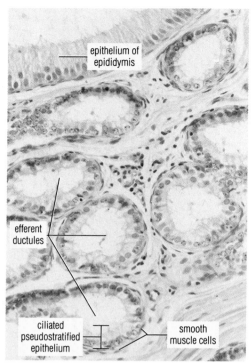

Figure 102a **Efferent ductules** connect the rete testes to the epididymis (prep. slide, sec., 250×). (Photo by D. Morton)

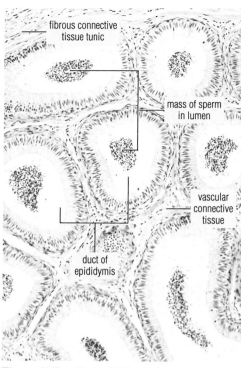

Figure 102b The **epididymis** functions to store sperm in its highly coiled duct (prep. slide, sec., 100×). (Photo by D. Morton)

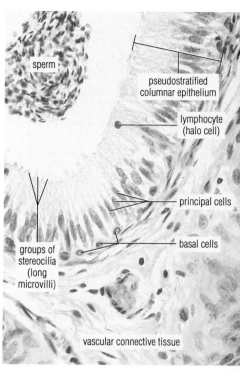

Figure 102c **Epithelium** of **epididymis** (prep. slide, sec., 400×). (Photo by D. Morton)

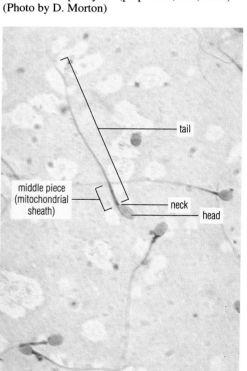

Figure 102d **Sperm** (prep. slide, sec., 1000×). (Photo by D. Morton)

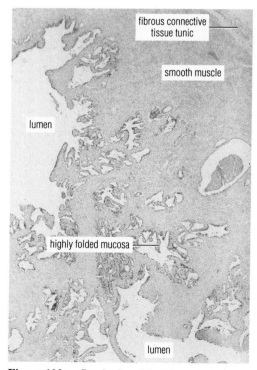

Figure 102e **Seminal vesicle**. The coiled lumen, bounded by a highly folded mucosa, appears several times in this typical section (prep. slide, sec., 25×). (Photo by D. Morton)

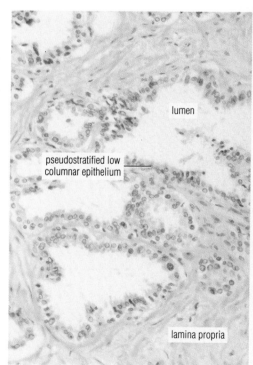

Figure 102f **Mucosa** of **seminal vesicle** (prep. slide, sec., 250×). (Photo by D. Morton)

Male/Prostate, Spermatic Duct, and Penis

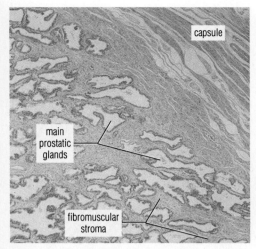

Figure 103a **Prostate** from a young man. Its glands drain into the prostatic urethra (prep. slide, sec., 25×). (Photo by D. Morton)

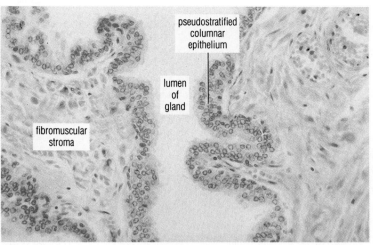

Figure 103b **Glandular epithelium** and **fibromuscular stroma** of **prostate** (prep. slide, sec., 250×). (Photo by D. Morton)

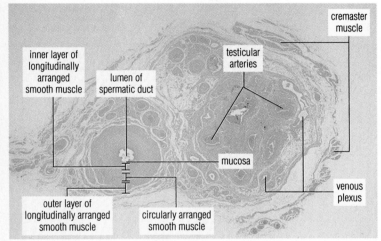

Figure 103c **Spermatic cord**, containing cross sections of **spermatic duct** (vas deferens) and **testicular arteries**. Note several vessels of the **venous plexus** (pampiniform plexus) and portions of the **cremaster muscle** (prep. slide, c.s., 8×). (Photo by D. Morton)

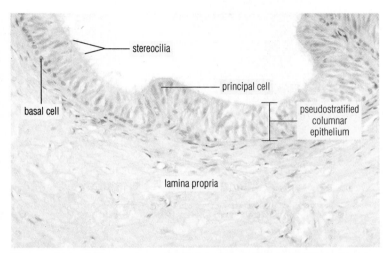

Figure 103d Mucosa of **spermatic duct** (prep. slide, c.s., 250×). (Photo by D. Morton)

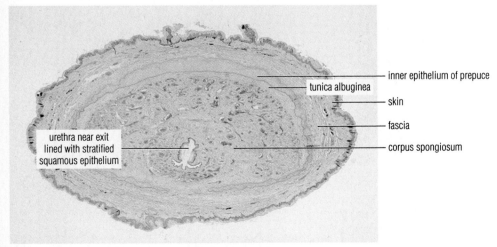

Figure 103e Cross section of **glans penis** of infant (prep. slide, c.s., 9×). (Photo by D. Morton)

104 REPRODUCTIVE SYSTEM *Female/Model and Ovary*

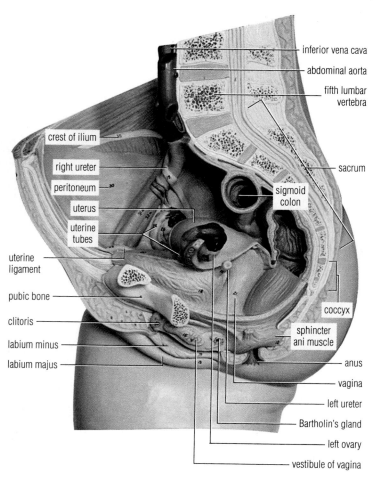

Figure 104a Model of **female reproductive system** and pelvis (0.30×). (Photo by D. Morton)

Figure 104b Model of **female reproductive system** with left half of uterus, urinary bladder, etc. removed (0.30×). (Photo by D. Morton)

Figure 104c Mammalian **ovary**. Arrow indicates direction of hilum (prep. slide, sec., 25×). (Photo by D. Morton)

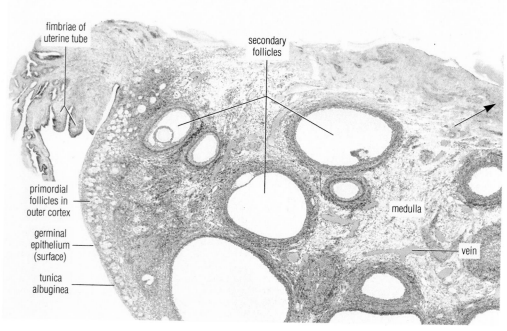

Female/Ovarian Structures — REPRODUCTIVE SYSTEM

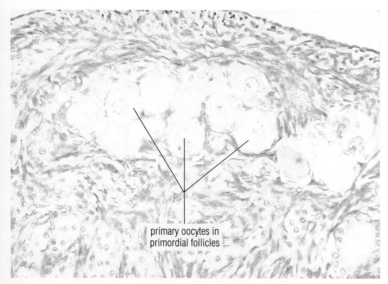

Figure 105a **Primordial follicles** (prep. slide, sec., 450×). (Photo by D. Morton)

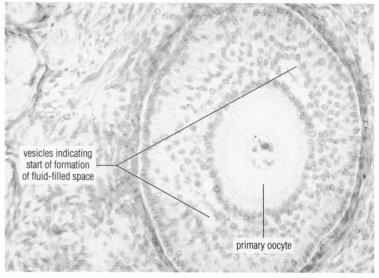

Figure 105b **Maturing primary follicle** (prep. slide, sec., 450×). (Photo by D. Morton)

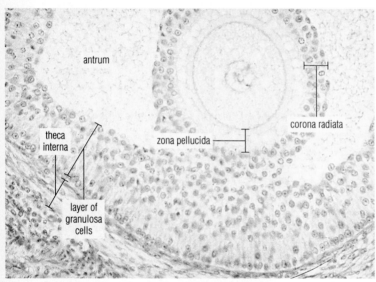

Figure 105c **Mature secondary follicle** (prep. slide, sec., 400×). (Photo by D. Morton)

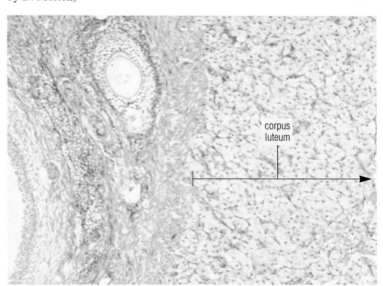

Figure 105d **Corpus luteum** (prep. slide, sec., 100×). (Photo by D. Morton)

Figure 105e **Corpus albicans** (prep. slide, sec., 100×). (Photo by D. Morton)

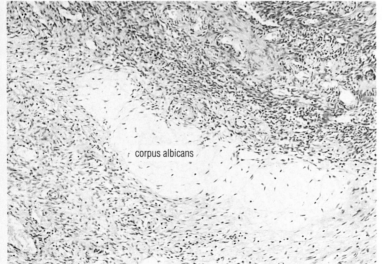

106 REPRODUCTIVE SYSTEM — Female/Uterine Tube and Uterus

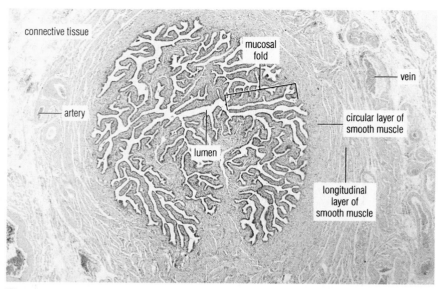

Figure 106a Ampulla region of **uterine tube** (oviduct) (prep. slide, c.s., 30×). (Photo by D. Morton)

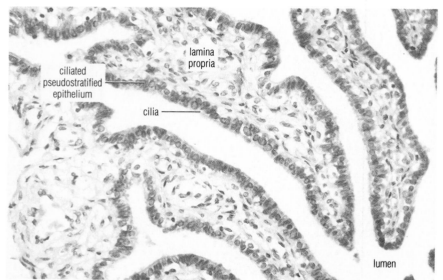

Figure 106b Mucosa of **uterine tube** (ampulla region) (prep. slide, c.s., 300×). (Photo by D. Morton)

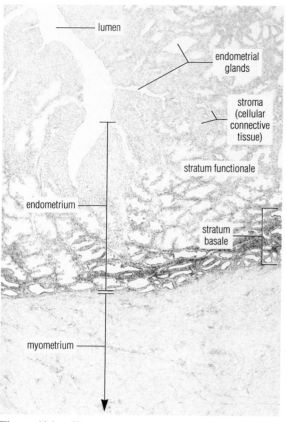

Figure 106c Uterus in **secretory phase** (prep. slide, sec., 30×). (Photo by D. Morton)

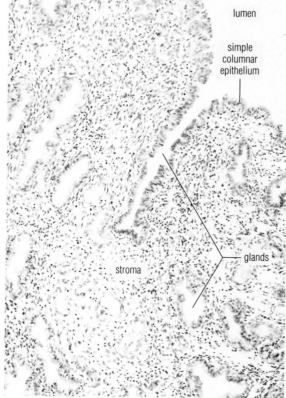

Figure 106d **Endometrium** (secretory phase) (prep. slide, sec., 100×). (Photo by D. Morton)

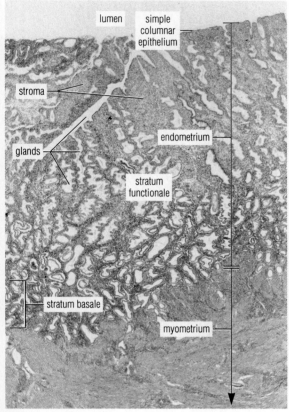

Figure 107a Endometrium (progravid phase) (prep. slide, sec., 30×). (Photo by D. Morton)

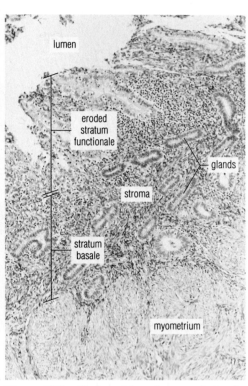

Figure 107b Endometrium (menstrual phase) (prep. slide, sec., 120×). (Photo by D. Morton)

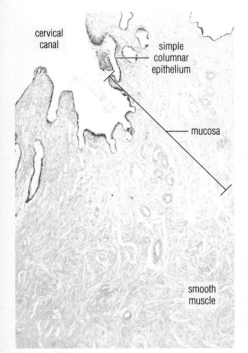

Figure 107c Cervix (uterine side) (prep. slide, l.s., 20×). (Photo by D. Morton)

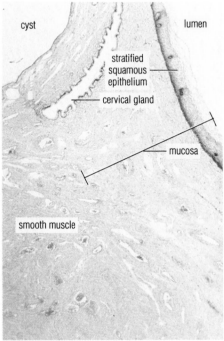

Figure 107d Cervix (vaginal side) (prep. slide, l.s., 20×). (Photo by D. Morton)

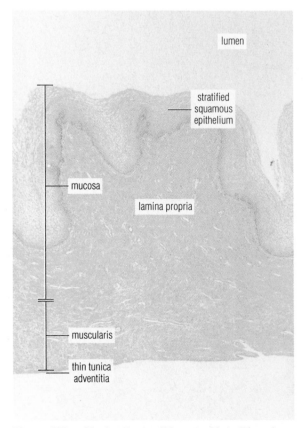

Figure 107e Vagina (prep. slide, sec., 30×). (Photo by D. Morton)

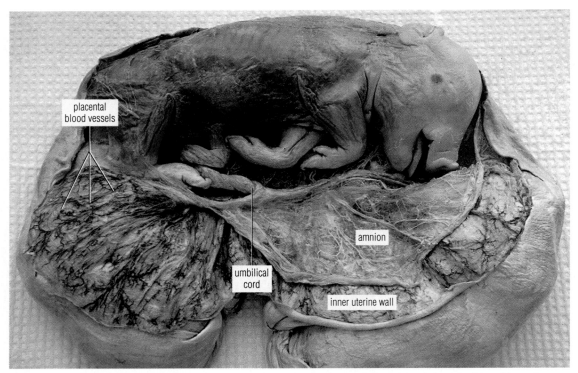

Figure 108a Extraembryonic membranes of the **fetal pig**. Fetal vessels have been injected with red latex for arteries and blue latex for veins (0.60×). (Specimen courtesy Wards Natural Science Establishment, Inc.; photo by D. Morton)

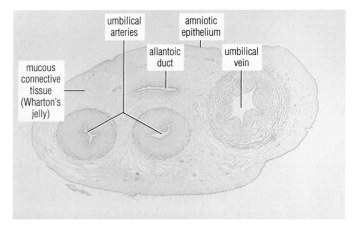

Figure 108b **Umbilical cord**, which functions to connect fetus and placenta (prep. slide, o.s., 10×). (Photo by D. Morton)

Figure 108c On the fetal side of the **placenta**, main-stem villi project from the chorionic plate. Smaller branches extend from the main stem. Branching continues down to the level of the terminal chorionic villi, which are suspended in the intervillous space (prep. slide, sec., 7×). (Photo by D. Morton)

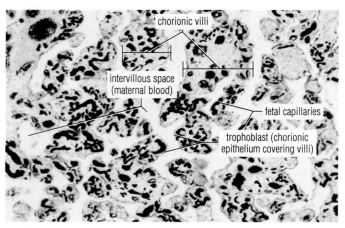

Figure 108d **Chorionic villi** consist of fetal capillaries surrounded by sparse connective tissue and covered by a mostly syncytial trophoblast. In life, they are surrounded by maternal blood in the **intervillous space**. In this placenta, the fetal blood vessels have been injected with carbon (prep. slide, sec., 100×). (Photo by D. Morton)

Female/Breast and Mammary Gland REPRODUCTIVE SYSTEM

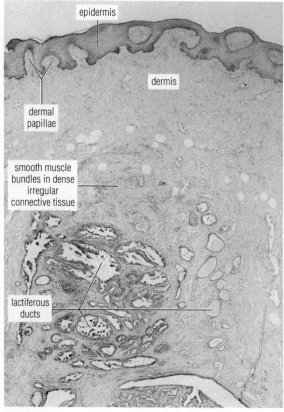

Figure 109a Nipple (prep. slide, sec., 30×). (Photo by D. Morton)

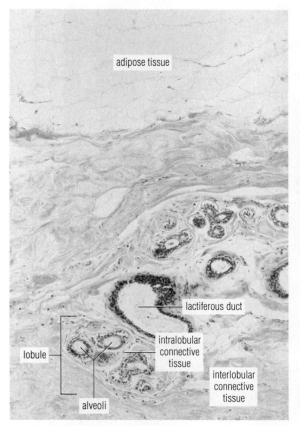

Figure 109b Inactive mammary gland (prep. slide, sec., 100×). (Photo by D. Morton)

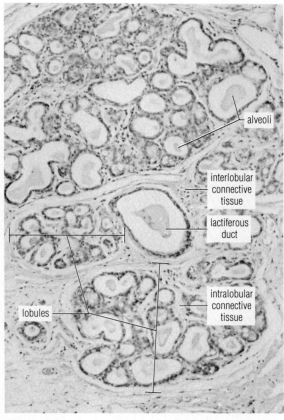

Figure 109c Active mammary gland (prep. slide, sec., 100×). (Photo by D. Morton)

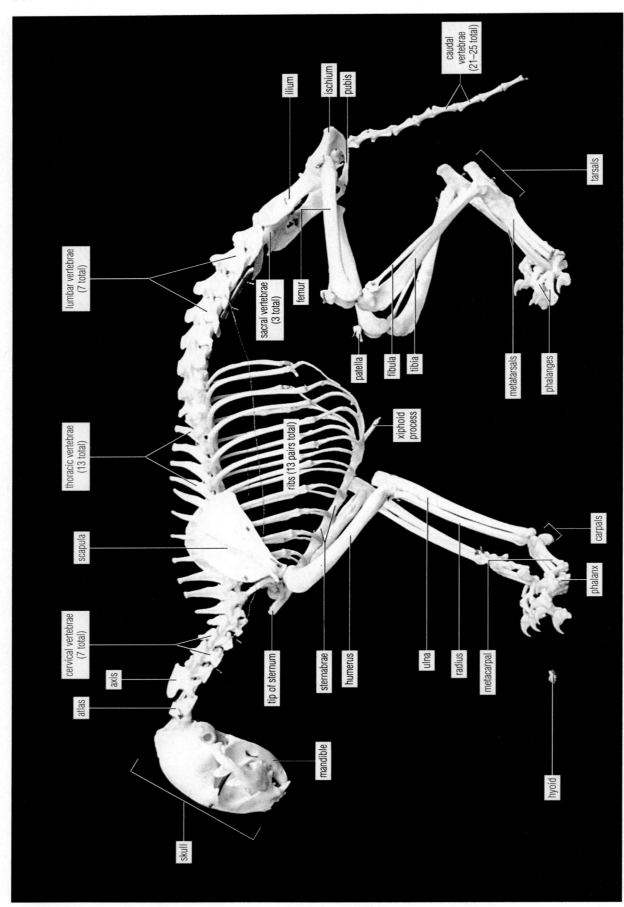

Figure 110a **Cat skeleton.** Differences from human skeleton include the numbers of thoracic, lumbar, sacral, and caudal (coccygeal) vertebrae, 13 rather than 12 ribs—9 true and 4 false, one of which is floating—an unfused sternum, fused scaphoid and lunate bones, claws, a reduced metatarsal in the back paw with no phalanges, and disarticulated clavicles as seen in Figure 114a (0.50×). (Photo by D. Morton)

CAT DISSECTION
Skinned Cat

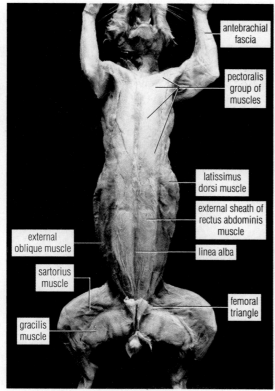

Figure 111a **Anterior view** of preserved **skinned** male **cat** with subcutaneous skeletal muscle and most of the superficial fascia removed. Deep fascia covers and binds together the undissected skeletal muscles (0.10×). (Photo by D. Morton)

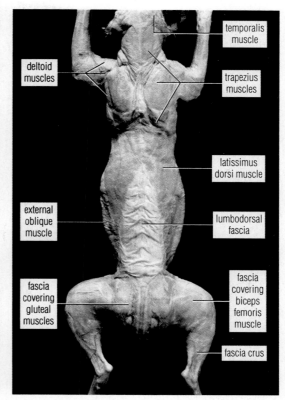

Figure 111b **Posterior view** of preserved **skinned** male **cat** (0.10×). (Photo by D. Morton)

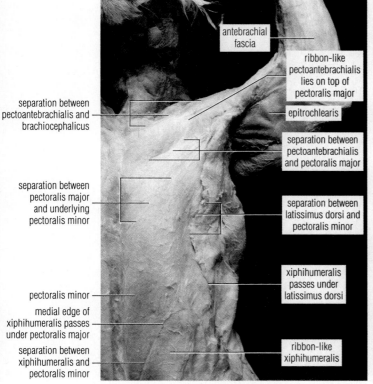

Figure 111c Chest and **medial surface** of the left **arm** with undissected skeletal muscles (0.40×). (Photo by D. Morton)

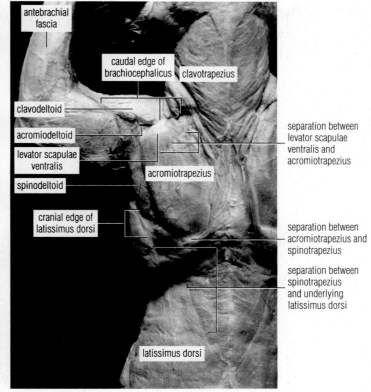

Figure 111d Upper back and **lateral surface** of the left **arm** with undissected skeletal muscles (0.40×). (Photo by D. Morton)

112　Cat Dissection　*Skeletal Muscles/Neck*

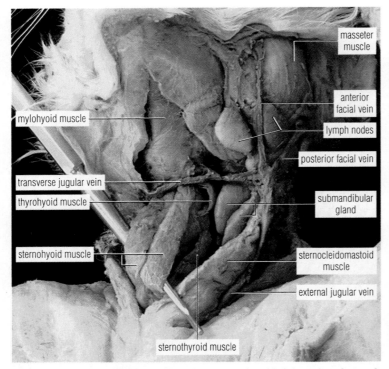

Figure 112a　Superficial muscles of the ventral and left lateral surfaces of **neck** (0.90×). (Photo by D. Morton)

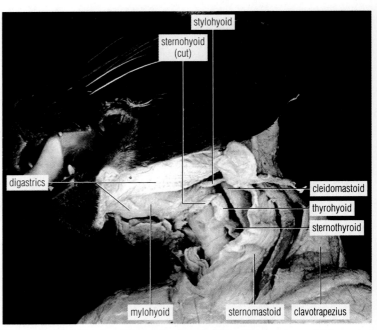

Figure 112b　Stylohyoid muscle of **neck**. Note that the sternocleidomastoid muscles of humans are separate in the cat (0.75×). (Photo by D. Morton)

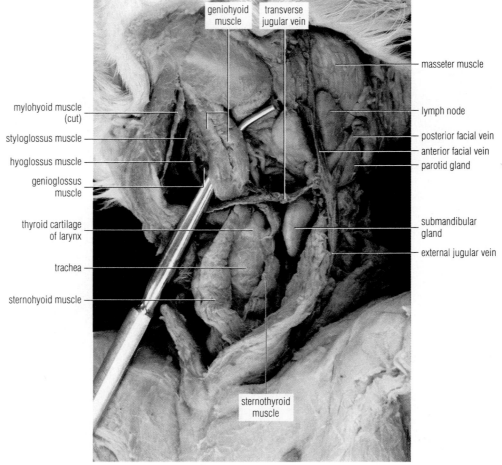

Figure 112c　Deeper muscles of the ventral **neck** (1×). (Photo by D. Morton)

Skeletal Muscles/Chest and Arm **CAT DISSECTION**

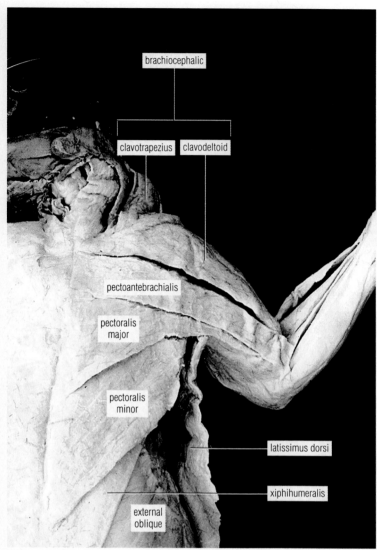

Figure 113a **Superficial muscles** of **chest**. Humans have neither pectoantebrachialis nor xiphihumeralis muscles, and our pectoralis major is larger and completely covers the pectoralis minor (0.70×). (Photo by D. Morton)

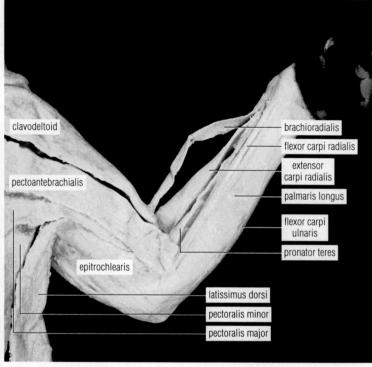

Figure 113b **Superficial muscles** of **medial arm**. Humans do not have an epitrochlearis muscle (0.70×). (Photo by D. Morton)

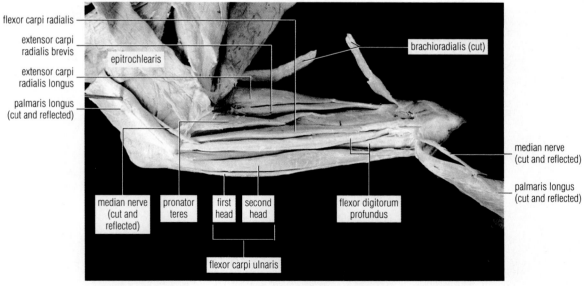

Figure 113c **Deeper muscles** of **medial forearm** (1.1×). (Photo by D. Morton)

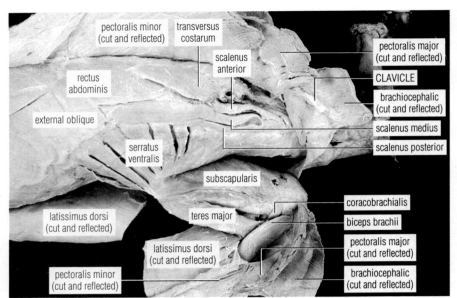

Figure 114a Deeper muscles of chest. A number of deep human neck muscles, including the scalenus muscles, are located primarily in the trunk of the cat. In the cat, the biceps brachii has only one head, with its origin on the supraglenoid tubercle of the scapula. Transversus costarum muscles are absent in humans (0.50×). (Photo by D. Morton)

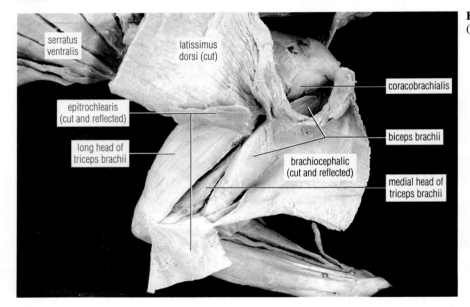

Figure 114b Deeper muscles of **medial forearm** (0.80×). (Photo by D. Morton)

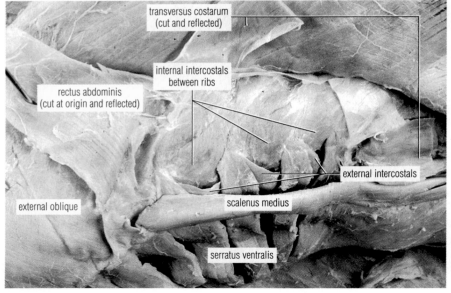

Figure 114c **Ventral view** of **intercostal muscles** (1×). (Photo by D. Morton)

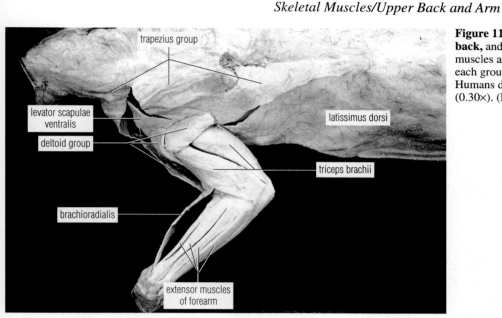

Figure 115a Superficial muscles of **dorsal neck, upper back,** and **lateral arm**. In the cat, there are three deltoid muscles and three trapezius muscles, and the most anterior of each group is fused to form the brachiocephalic muscle. Humans do not have levator scapulae ventralis muscles (0.30×). (Photo by D. Morton)

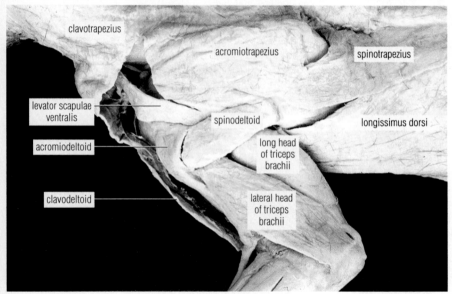

Figure 115b Superficial muscles of **upper back** and **lateral arm**. Clavodeltoid muscles are absent in humans, and the acromio- and spinodeltoids are fused into deltoid muscles. Also, in humans, the clavo-, acromio-, and spinotrapezius muscles are fused into the trapezius muscle (0.60×). (Photo by D. Morton)

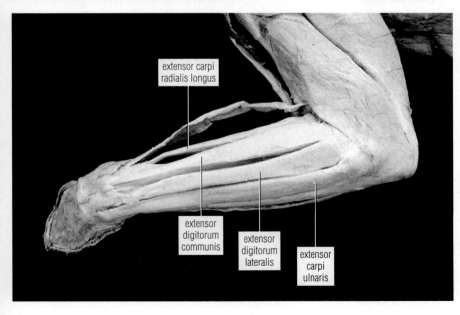

Figure 115c Superficial muscles of **lateral forearm**. Humans do not have extensor digitorum lateralis muscles. In the cat, the superficial supinator muscle of humans is located under the extensor digitorum lateralis muscle (1×). (Photo by D. Morton)

Cat Dissection — Skeletal Muscles/Upper Back and Arm

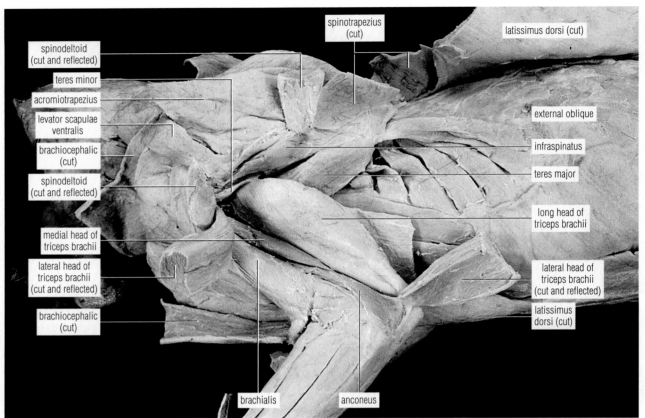

Figure 116a Deeper muscles of upper back and lateral arm (0.40×). (Photo by D. Morton)

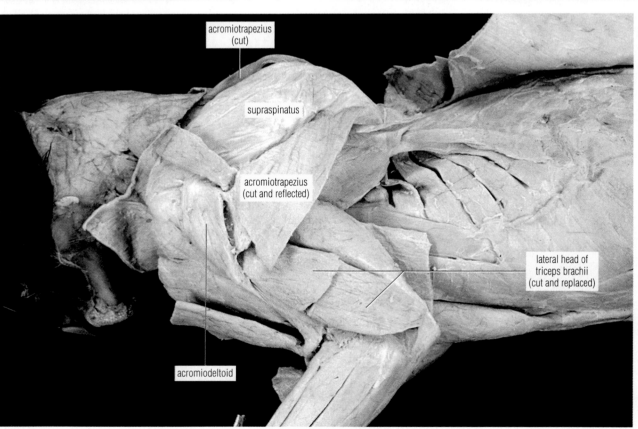

Figure 116b Supraspinatus muscle (0.40×). (Photo by D. Morton)

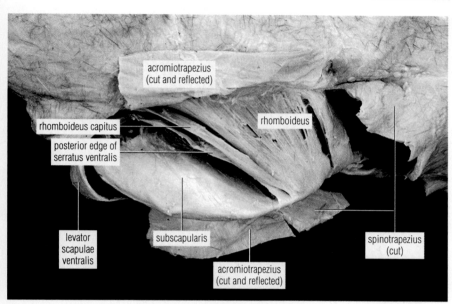

Figure 117a In this **dorsal view** of the **deeper muscles** of the **upper back**, the scapula is pulled away from the vertebral column to reveal the subscapularis muscle and the rhomboideus muscles. Humans do not have rhomboideus capitus muscles (0.70×). (Photo by D. Morton)

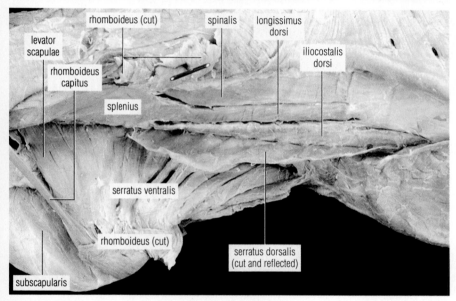

Figure 117b **Deepest muscles** of **upper back**. The cranial portions of the serratus ventralis muscles of the cat, which arise from the cervical vertebrae, are homologous to the levator scapulae muscles of humans. Compared to the situation in the cat, humans have widely separated serratus anterior and posterior muscles (0.70×). (Photo by D. Morton)

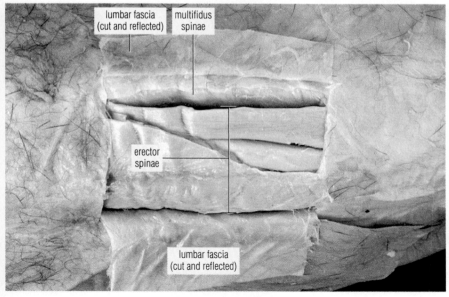

Figure 117c Window cut into the lumbodorsal fascia of the **lower back** (1.2×). (Photo by D. Morton)

118 CAT DISSECTION — Skeletal Muscles/Abdomen

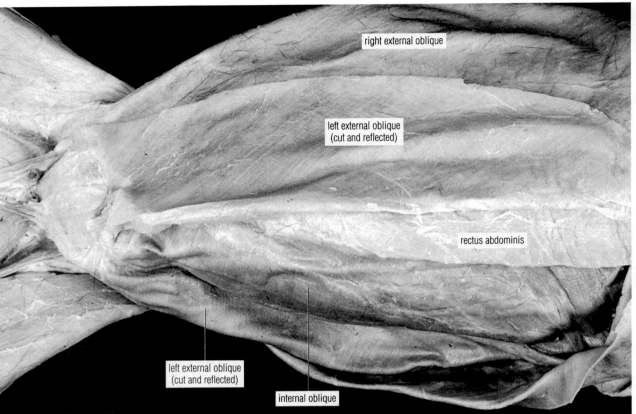

Figure 118a Abdominal muscles (1×). (Photo by D. Morton)

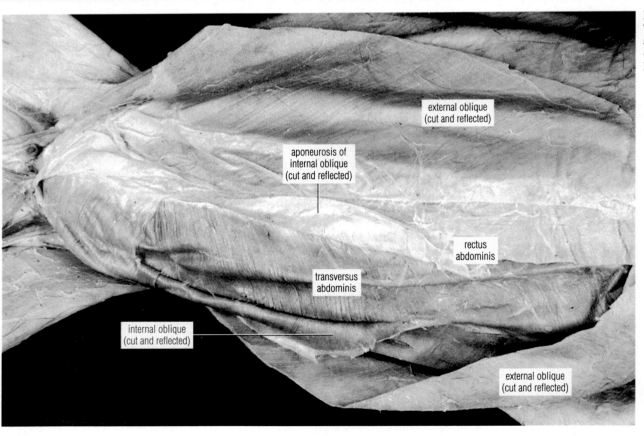

Figure 118b Transversus abdominis muscle (1×). (Photo by D. Morton)

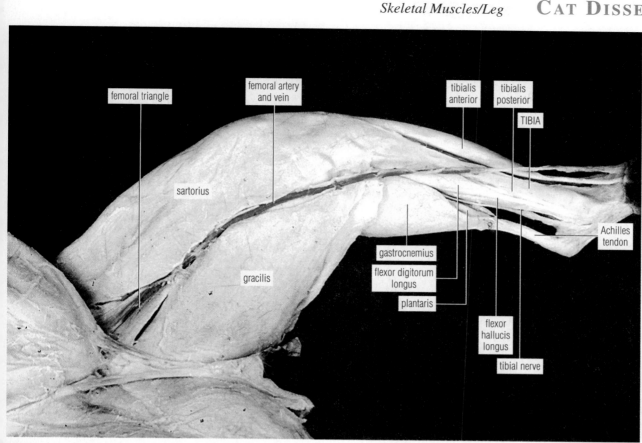

**Figure 119a
Superficial muscles** of **medial leg** (0.70×).
(Photo by D. Morton)

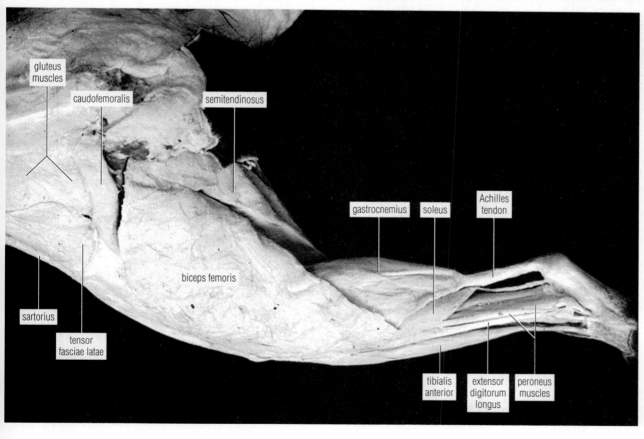

**Figure 119b
Superficial muscles** of **lateral leg**.
Humans do not have caudofemoralis muscles (0.70×).
(Photo by D. Morton)

Cat Dissection — Skeletal Muscles/Thigh and Lower Leg

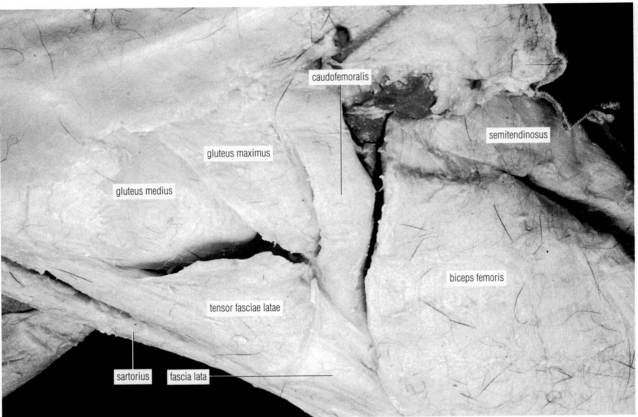

Figure 120a Superficial muscles of lateral thigh. The gluteus maximus muscle of humans is larger and almost completely covers the gluteus medius (1.7×). (Photo by D. Morton)

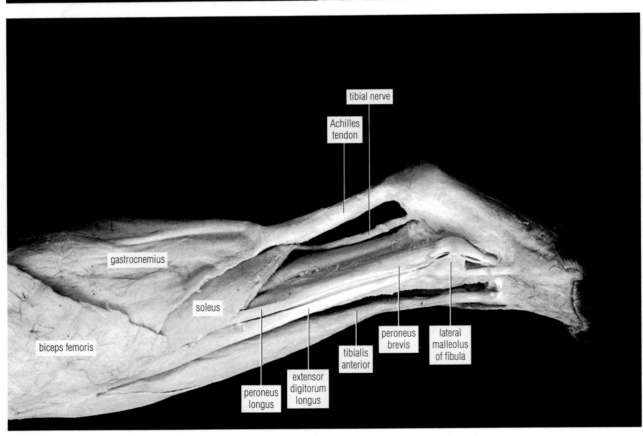

Figure 120b Lateral lower leg muscles (1.3×). (Photo by D. Morton)

Skeletal Muscles/Thigh CAT DISSECTION

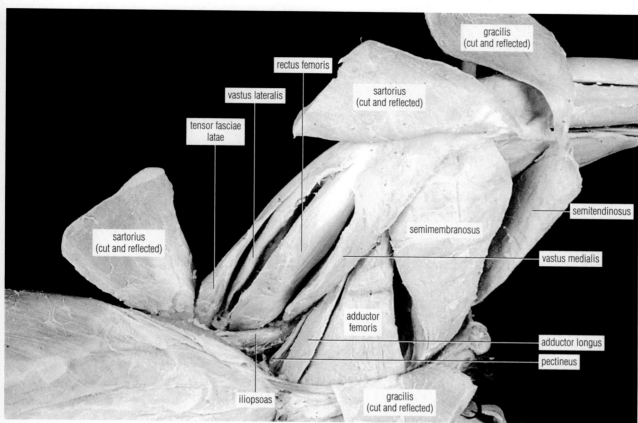

Figure 121a Deeper muscles of medial thigh. In humans the adductor femoris muscle of the cat is divided into the adductor brevis and the adductor magnus. There are separate iliacus and psoas major muscles in humans (0.80×). (Photo by D. Morton)

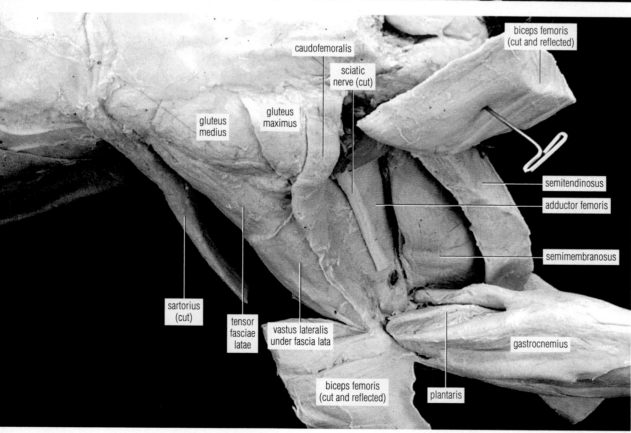

Figure 121b Deeper muscles of lateral thigh (1×). (Photo by D. Morton)

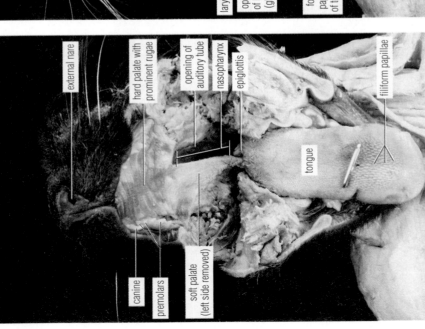

Figure 122a Oral cavity. The spinelike filiform papillae of cats are used for grooming. Rugae, horizontal ridges in the hard palate, are less prominent in humans. The first maxillary premolar and the molar of cats are reduced in size, and the third maxillary premolar and mandibular molar form shearing or *carnassial* teeth (0.80×). (Photo by D. Morton)

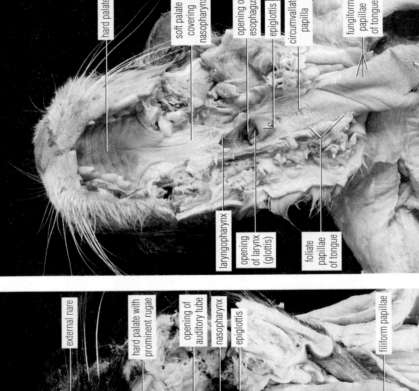

Figure 122b Pharynx. The sides of the fauces (the space between the oral cavity and the pharynx) are cut to open the oropharynx and the epiglottis pinned anteriorly to reveal the laryngopharynx and the opening into the nasopharynx. Humans do not have foliate papillae (0.70×). (Photo by D. Morton)

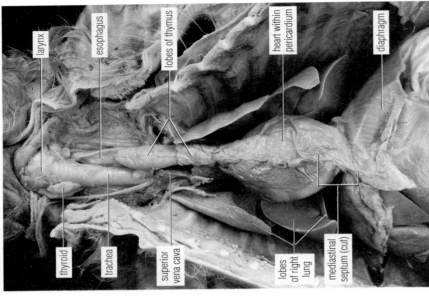

Figure 122c Thoracic cavity. (0.90×). (Photo by D. Morton)

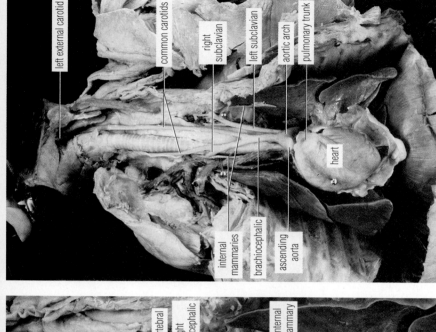

Figure 123a Respiratory and circulatory systems in thoracic cavity. The human left lung has only two lobes (0.50×). (Photo by D. Morton)

Figure 123b Thoracic veins. In humans, the vertebral and internal jugular veins drain the brain, and the external jugular drains the outside regions of the head. The main vein draining almost all of the head in the cat is the external jugular. The delicate internal jugulars are often not injected or lost during dissection. Humans do not have a common jugular vein or a transverse jugular connecting the external jugular veins (1.0×). (Photo by D. Morton)

Figure 123c Thoracic arteries. In the cat, the internal carotids are much reduced in size, and blood flows to the brain via the external carotids to the *rete mirabile*, a fine network of arteries that are intertwined with the orbital venous sinus, which drains blood from the nasal region. This countercurrent system—flow in opposite directions—cools the blood going to the brain and is common in mammals that, like cats, suffer rises in body temperature due to a combination of heavy fur and bursts of intense activity or that need to conserve water in a hot, dry environment. In humans, arterial blood reaches the brain through both the vertebral and internal carotid arteries (0.45×). (Photo by D. Morton)

124 CAT DISSECTION — Heart/Arteries

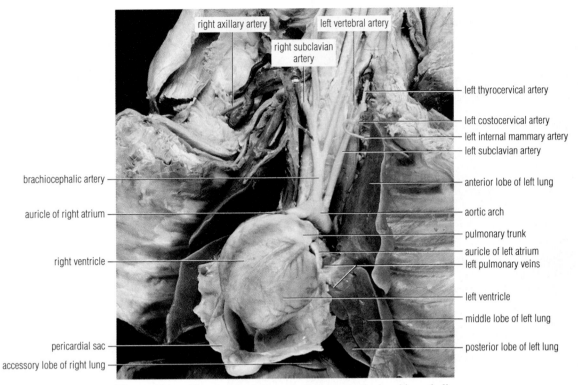

Figure 124a **Heart** and **aortic arch**. In humans, the brachiocephalic artery branches into the right subclavian and right common carotid arteries, and the left common carotid branches directly from the aortic arch (0.65×). (Photo by D. Morton)

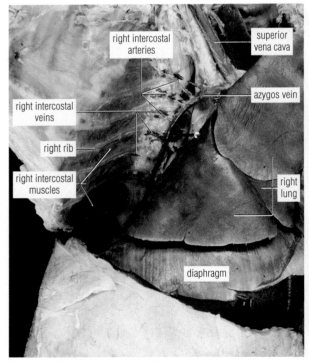

Figure 124b **Azygos and intercostal veins** draining the right and left intercostal muscles. Humans also have a hemiazygos vein that drains the left intercostal muscles and empties into the azygos vein (0.65×). (Photo by D. Morton)

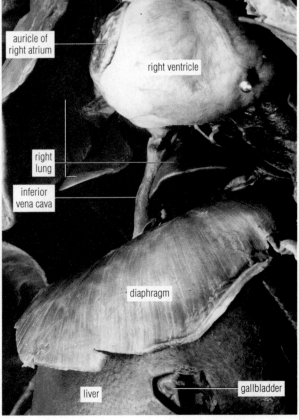

Figure 124c **Inferior vena cava** (1.4×). (Photo by D. Morton)

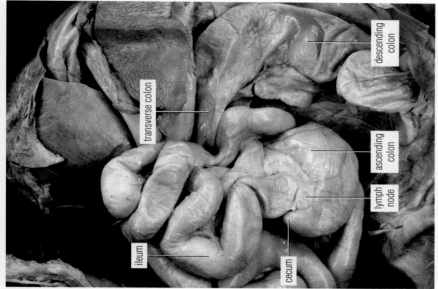

Figure 125c Pushing the small intestine to the right uncovers the **large intestine**. Humans have an appendix arising from the cecum; cats do not (1.1×). (Photo by D. Morton)

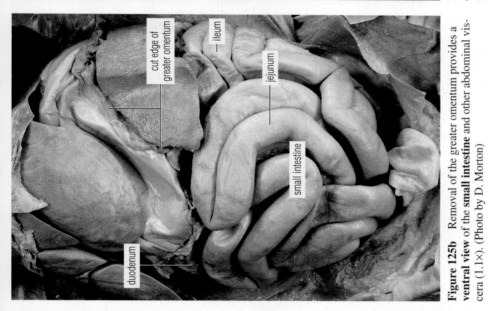

Figure 125b Removal of the greater omentum provides a **ventral view** of the **small intestine** and other abdominal viscera (1.1×). (Photo by D. Morton)

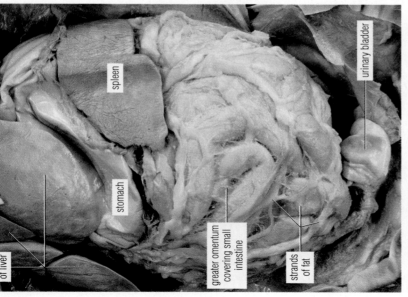

Figure 125a Abdominal cavity with **greater omentum** covering the small intestine (1×). (Photo by D. Morton)

Cat Dissection — Abdominal Cavity/Digestive System

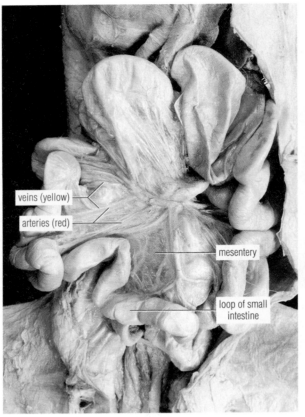

Figure 126a A dorsal **mesentery** supports the small intestine (1×). (Photo by D. Morton)

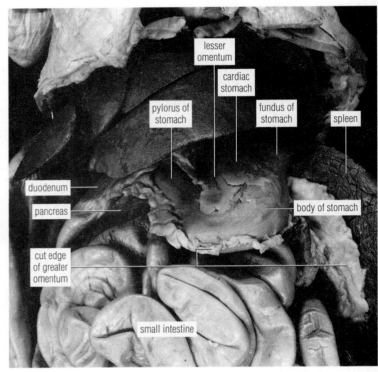

Figure 126b **Stomach** (0.65×). (Photo by D. Morton)

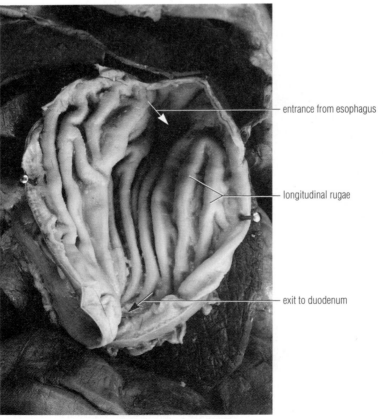

Figure 126c **Internal surface** of **stomach** (1.1×). (Photo by D. Morton)

Abdominal Cavity/Digestive System   **CAT DISSECTION**

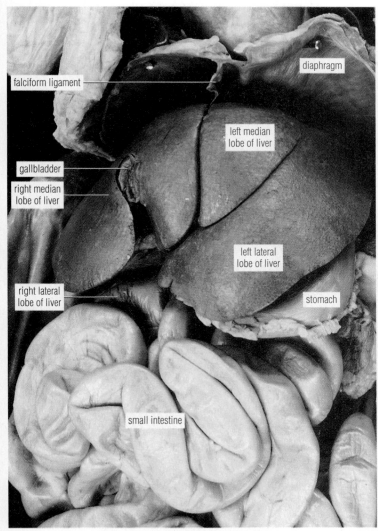

Figure 127a   **Ventral view** of **liver**. Humans have four liver lobes rather than the cat's five. The fifth lobe is visible in Figure 129a (0.80×). (Photo by D. Morton)

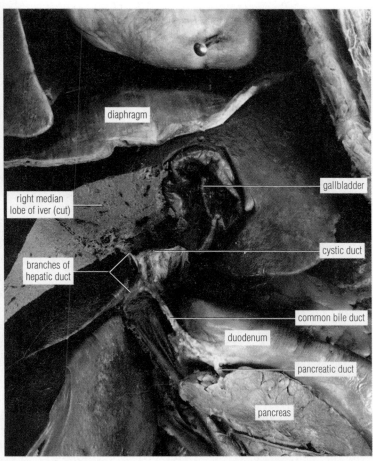

Figure 127b   **Gallbladder**, **pancreas**, and associated ducts. The cat pancreas is located between the stomach and the first loop of the duodenum. In humans, the pancreas extends from the duodenum horizontally along the dorsal body wall next to the greater curvature of the stomach, and an accessory pancreatic duct is present (1.8×). (Photo by D. Morton)

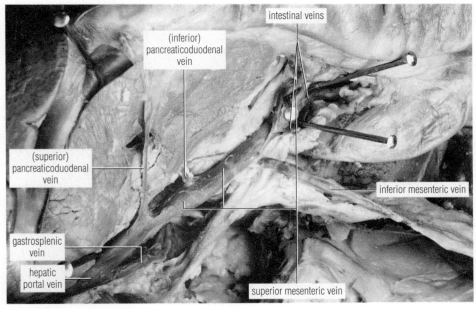

Figure 127c   **Hepatic portal system** (1.6×). (Photo by D. Morton)

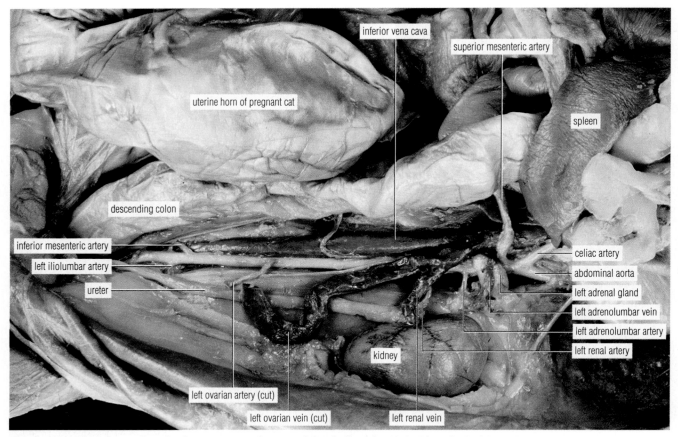

Figure 128a Blood vessels and urinary system of upper abdominal cavity (1×). (Photo by D. Morton)

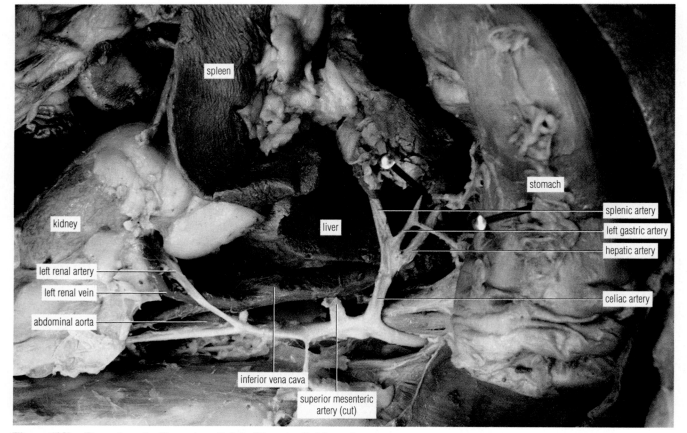

Figure 128b Branches of celiac artery (1.6×). (Photo by D. Morton)

Urinary System   CAT DISSECTION   129

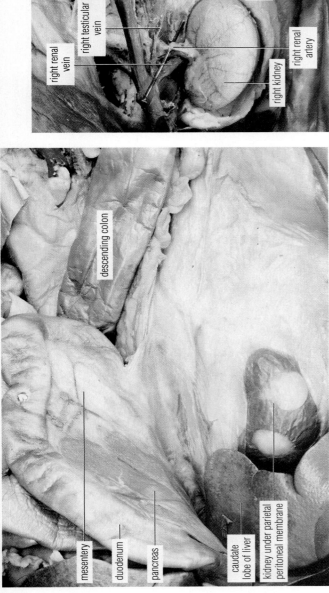

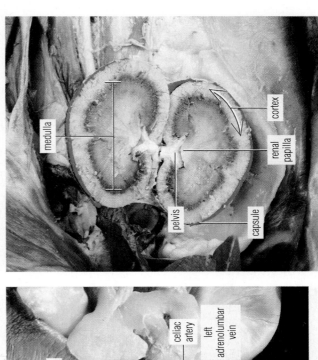

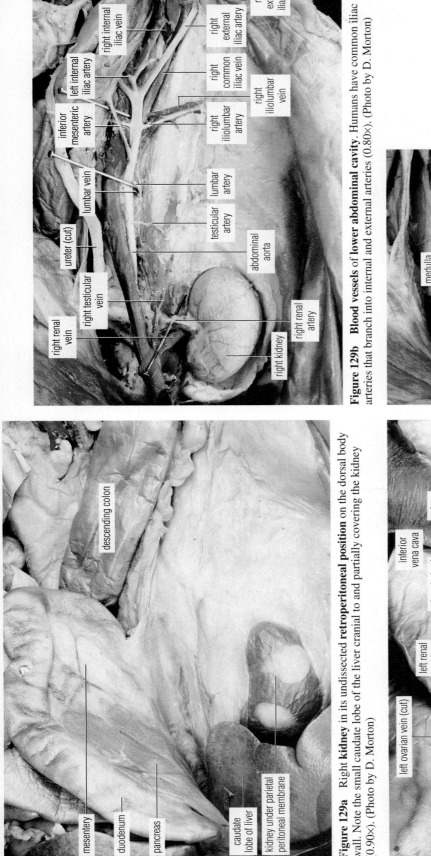

Figure 129a Right kidney in its undissected retroperitoneal position on the dorsal body wall. Note the small caudate lobe of the liver cranial to and partially covering the kidney (0.90×). (Photo by D. Morton)

Figure 129b Blood vessels of lower abdominal cavity. Humans have common iliac arteries that branch into internal and external arteries (0.80×). (Photo by D. Morton)

Figure 129c Dorsal view of left kidney (1.1×). (Photo by D. Morton)

Figure 129d Coronal (frontal) section of right kidney. Humans have nine or so renal papillae compared to the cat's one (0.80×). (Photo by D. Morton)

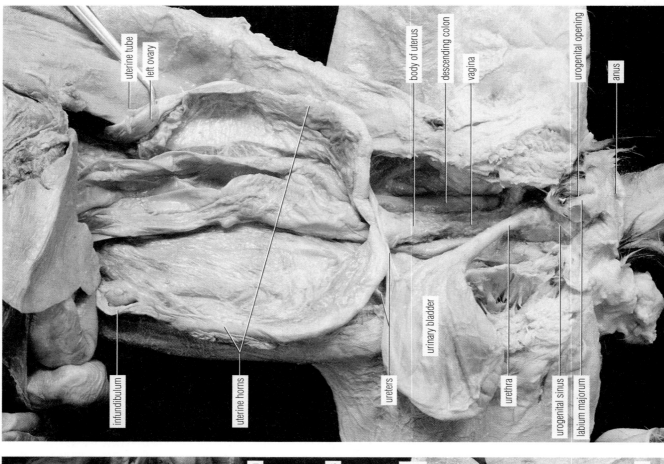

Figure 130a **Male reproductive system** and **lower digestive and urinary systems.** Male humans have an additional pair of sex accessory glands, the seminal vesicles, which drain into the spermatic duct at its base. Also the penis is not supported by a bone, (os penis) as is the case in the cat. The human urinary bladder does not have a short neck connecting it to the urethra (0.90×). (Photo by D. Morton)

Figure 130b **Female reproductive system.** Female humans have a simplex uterus rather than the bicornuate uterus of cats, and longer and more prominent uterine tubes. Also, humans have an oval depression (vestibule) at the point where the urethra and vagina open to the outside of the body rather than the tubular urogenital sinus of cats (1×). (Photo by D. Morton)

Reproductive System/Pregnant Female **Cat Dissection**

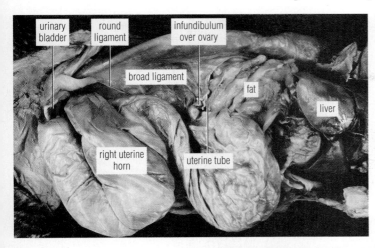

Figure 131a **Upper reproductive system** of a **pregnant** cat (0.50×). (Photo by D. Morton)

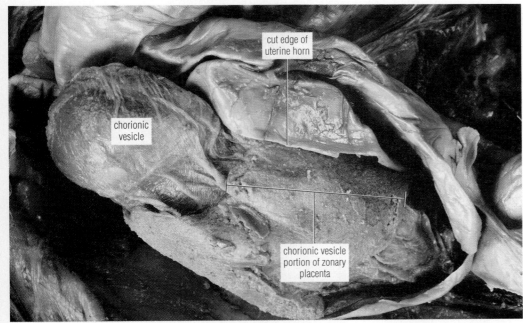

Figure 131b **Chorionic vesicle**. Humans have a discoidal (disk-shaped) placenta rather than the bandlike zonary placenta of cats (1.1×). (Photo by D. Morton)

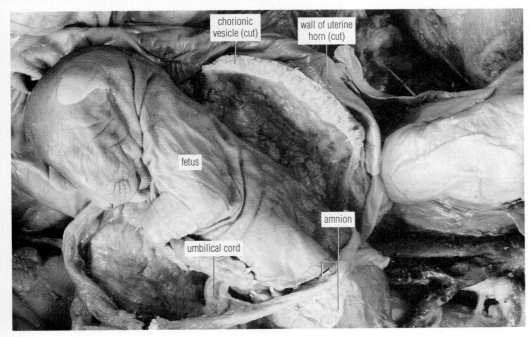

Figure 131c **Fetus**, **amnion**, and **umbilical cord** (1.2×). (Photo by D. Morton)

Fetal Pig Dissection — External Anatomy/Oral Cavity

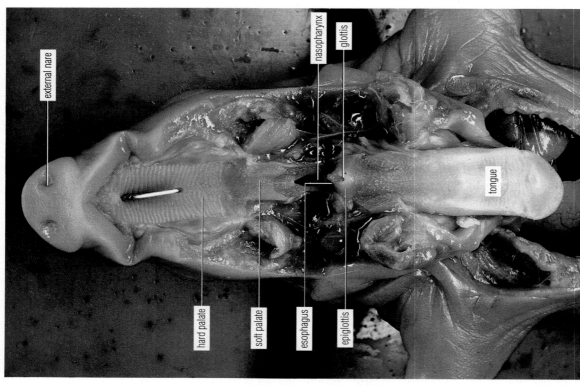

Figure 132c Oral cavity and pharynx. The opening of the esophagus can't be seen but is just behind the glottis (1×). (Photo by D. Morton)

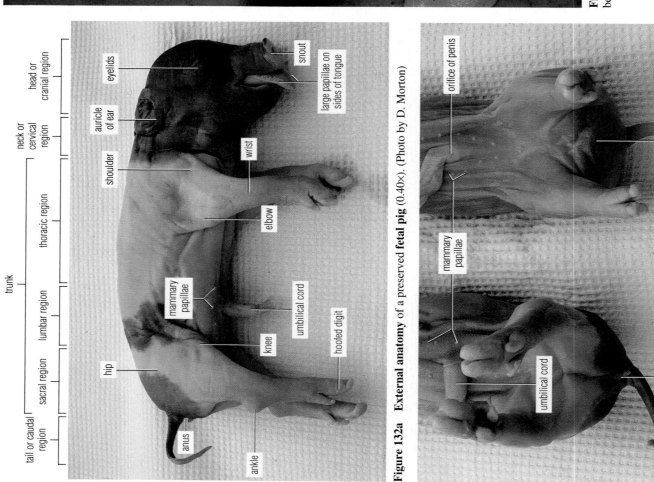

Figure 132a External anatomy of a preserved fetal pig (0.40×). (Photo by D. Morton)

Figure 132b Ventral view of female (left) and male (right) fetal pigs (0.40×). (Photo by D. Morton)

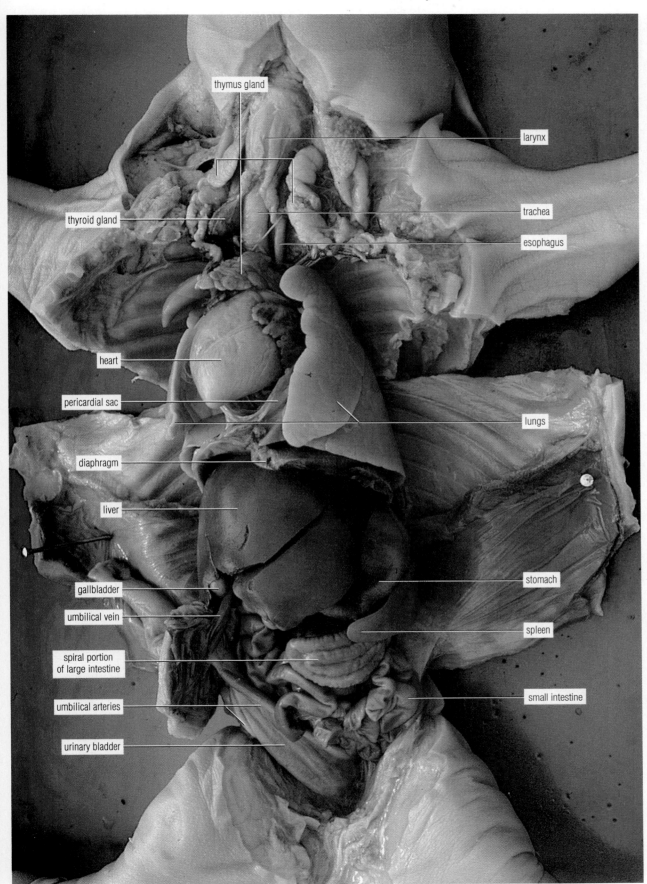

Figure 133a **Ventral view** of **internal anatomy** (1.3×). (Photo by D. Morton)

134 FETAL PIG DISSECTION *Circulatory System/Veins and Heart*

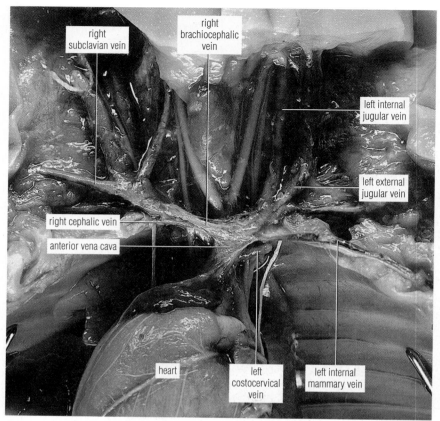

Figure 134a **Ventral view** of **thoracic veins** (2.5×). (Photo by D. Morton)

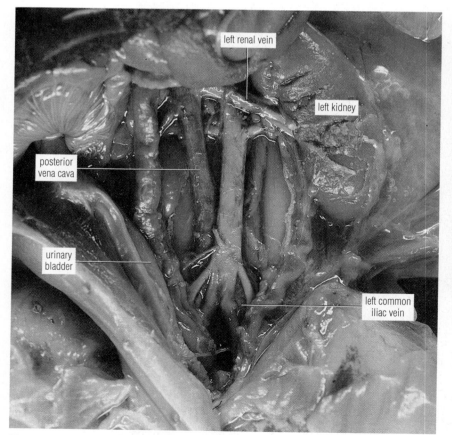

Figure 134b **Ventral view** of **abdominopelvic veins** (3×). (Photo by D. Morton)

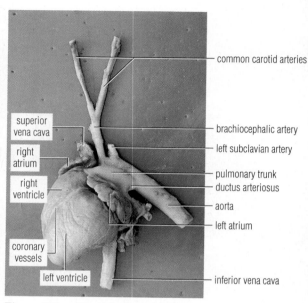

Figure 134c **Ventral view** of fetal pig **heart** (1.1×). (Photo by D. Morton)

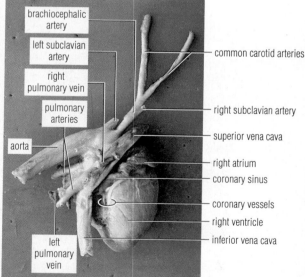

Figure 134d **Dorsal view** of fetal pig **heart** (1.1×). (Photo by D. Morton)

Circulatory System/Arteries — FETAL PIG DISSECTION

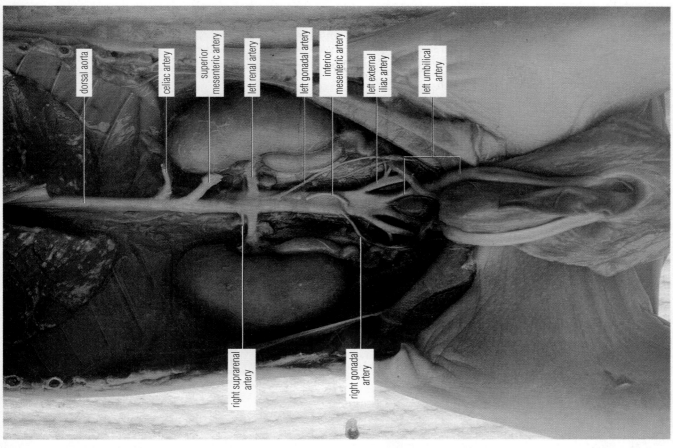

Figure 135b Ventral view of abdominopelvic arteries (2.2×). (Photo by D. Morton)

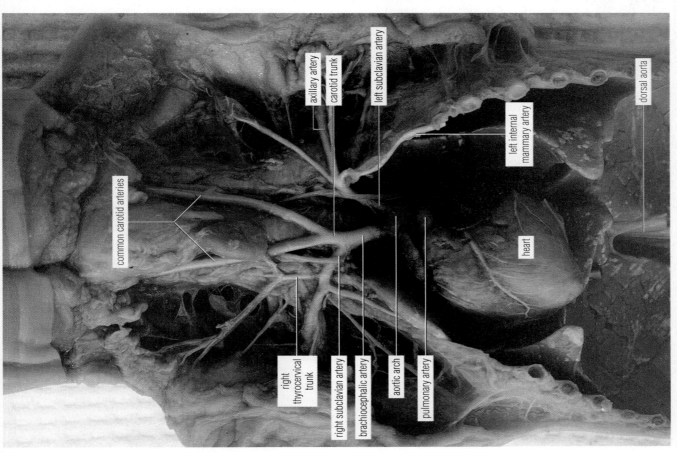

Figure 135a Ventral view of thoracic arteries (2.3×). (Photo by D. Morton)

136 FETAL PIG DISSECTION — Urinary and Reproductive Systems

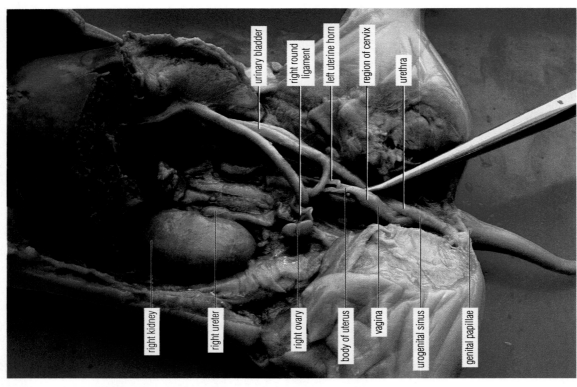

Figure 136a Ventral view of the urogenital system of the **male fetal pig** (1×). (Photo by D. Morton)

Figure 136b Ventral view of the urogenital system of the **female fetal pig** (1×). (Photo by D. Morton)

Nervous System **FETAL PIG DISSECTION** 137

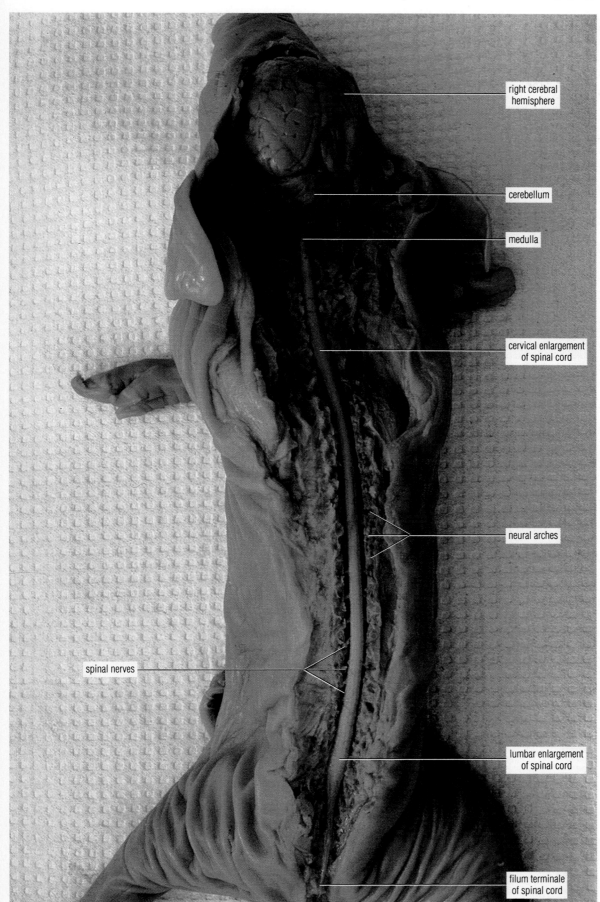

Figure 137a Dorsal view of **central nervous system** (0.80×). (Photo by D. Morton)

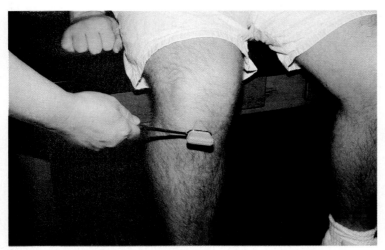

Figure 138a Patella reflex. Tapping the patella ligament stetches the quadriceps femoris muscles and their muscle spindles. This activates a monosynaptic reflex arc, resulting in contraction of these muscles and extension of the lower leg at the knee (0.20×). (Photo by D. Morton)

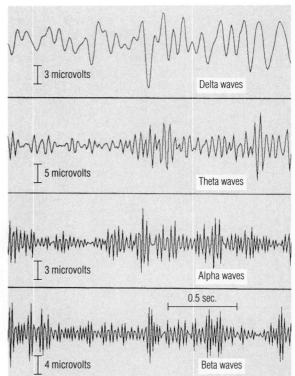

Figure 138b Electroencephalogram (EEG). (Photo by D. Morton)

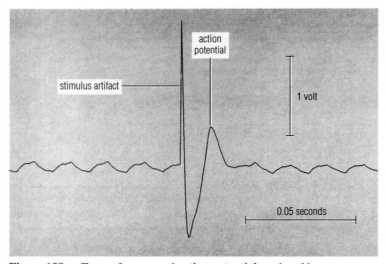

Figure 138c Trace of compound **action potential** produced by a submaximal stimulus delivered to the spinal cord end of a frog sciatic nerve and received by recording electrodes placed toward the knee. (Photo by D. Morton)

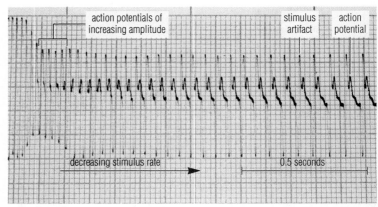

Figure 138d A high rate of submaximal stimuli (first four stimuli from the left) produces no action potentials. This is because the frog sciatic nerve is in the **absolute refractory period** (up to about 0.02 seconds for this trace). Further decrease in the stimulus rate (the next five stimuli) results in action potentials of increasing amplitude that finally plateau. The time between the first action potential produced and the start of the plateau (about 0.01 seconds in duration for this trace) is the **relative refractory period**. (Photo by D. Morton)

Doubly Pithing a Frog — PHYSIOLOGY

Figure 139a Step 1 Grasp the frog with your nondominant hand and bend its head forward between the index finger and second finger. With the nail on the index finger of the dominant hand, feel for the foramen magnum. (0.70×). (Photo by D. Morton)

Figure 139b Step 2 When it is located, press to leave a crease in the skin to mark its location (arrow). Place the tip of a dissecting needle in the center of the crease. (0.70×). (Photo by D. Morton)

Figure 139c Step 3 Push the needle into the cranial cavity and scramble the brain. The frog is now singly pithed. (0.70×). (Photo by D. Morton)

Figure 139d Step 4 To doubly pith the frog, withdraw the needle and reverse its direction. (0.70×). (Photo by D. Morton)

Figure 139e Step 5 Push the needle down the entire length of the vertebral canal to scramble the spinal cord. Until now, the frog's legs have been flexed at the hip, knee, and ankle. (0.70×). (Photo by D. Morton)

Figure 139f Step 6 A successful double-pith will result in a rigid extension of the legs, which will last for several seconds (0.40×). (Photo by D. Morton)

140 PHYSIOLOGY *Skeletal Muscles/Electromyography*

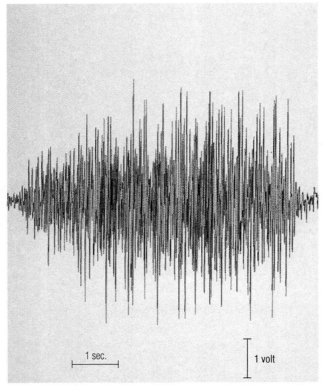

Figure 140a Typical electromyograph (EMG) from an isometric contraction of the biceps brachii. (Photo by D. Morton)

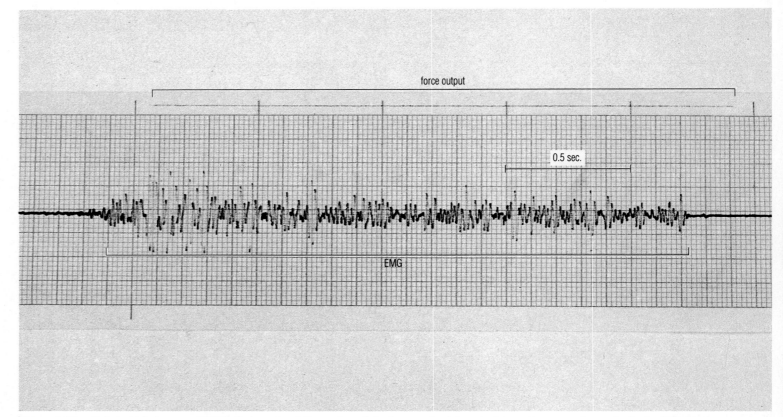

Figure 140b Relationship between **EMG** of triceps brachii muscle and the **force output** from its isometric contraction as the back of the hand rests on a pressure pad. Note force output starts and stops after the EMG starts and stops, respectively. (Photo by D. Morton)

Skeletal Muscles/Twitch PHYSIOLOGY

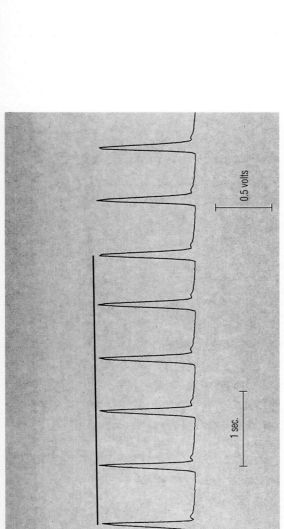

Figure 141a Treppe. Muscle twitches produced by a "cold" frog gastrocnemius muscle preparation activated by a series of submaximal stimuli delivered to the sciatic nerve. The line above the first six twitches highlights the slight increase in the peak force of contraction until the muscle "warms up." (Photo by D. Morton)

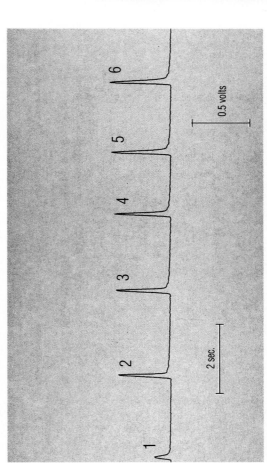

Figure 141b Recruitment of motor units. Stimuli of increasing strength applied to the sciatic nerve of a frog produced these six gastrocnemius muscle twitches. The minimal or liminal stimulus is the strength of stimulus that results in a muscle twitch (1). Increasing the stimulus strength returns a corresponding increase in the peak force of contraction (twitches 1 to 5) until it reaches a plateau (twitches 5 and 6). The maximal stimulus is the lowest strength that produces a maximum peak force of contraction. Stimulus strengths that lie between minimal and maximal are submaximal, and those above maximal are supramaximal. (Photo by D. Morton)

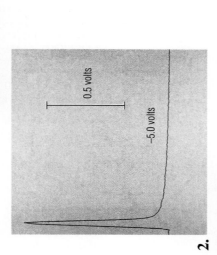

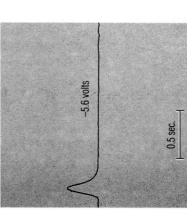

Figure 141c Muscle length versus **peak force of contraction.** Stretching a frog gastrocnemius muscle yields muscle length changes in the peak force of contraction of twitches resulting from stimuli of equal intensity. At relatively short muscle lengths (first twitch), the amplitude is low due to crowding in the sarcomeres. Stretching the muscle to near its ideal length produces a twitch close to maximum size (second twitch). Above this ideal length, the decreasing region of overlap between the thick and thin filaments in the sarcomeres again decreases the active force of contraction (third twitch). Note also that stretching the muscle increases the passive force generated by its elastic connective tissue components, as illustrated by higher baseline voltages. (Photo by D. Morton)

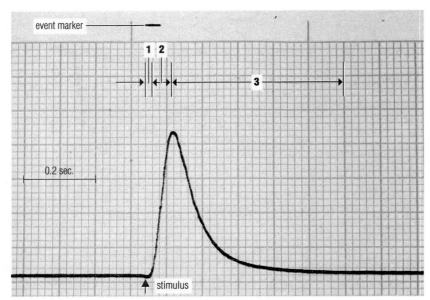

Figure 142a Time course of a **muscle twitch**. Muscle twitch produced by a frog sciatic nerve/gastrocnemius muscle preparation. The time between the stimulus (arrow at bottom) and the rise of the muscle twitch is the latent period (0.02 seconds in this trace). The latent period (1) encompasses those events involved with the conveying the "message to contract" to the level of the filaments. The rise time (0.06 seconds in this trace) of the muscle twitch, from the baseline to the peak force of contraction, is the contraction period (2). The fall time (0.5 seconds in this trace) is the relaxation period (3). (Photo by D. Morton)

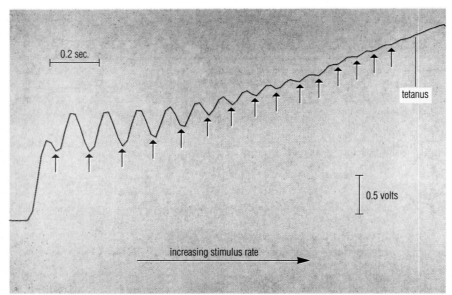

Figure 142b Twitch fusion, summation, and tetanus. Results obtained from a frog sciatic nerve/gastrocnemius muscle preparation when the stimulus rate increases over time. Twitches fuse, with each subsequent twitch adding on earlier and earlier to the previous one (arrows). This results in an increasing peak force of contraction (summation) and finally a straight line (tetanus). (Photo by D. Morton)

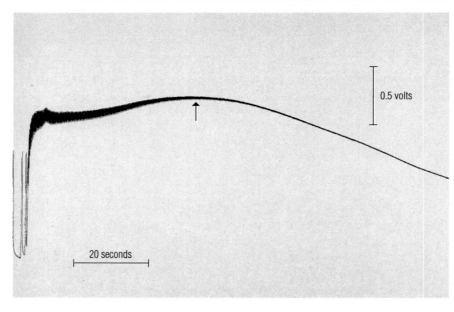

Figure 142c Fatigue. Results obtained from a frog sciatic nerve/gastrocnemius muscle preparation stimulated at a high rate. The force of contraction decreases (arrow) when replenishment of metabolic fuels cannot keep up with demand and sometimes ion imbalances occur. (Photo by D. Morton)

Circulatory System/Blood Pressure, Peripheral Blood Flow, and ECG

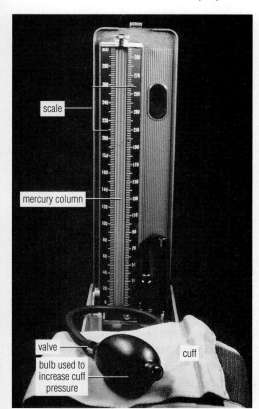

Figure 143a Sphygmomanometer. This instrument for estimating blood pressure in the brachial artery comes in many types. This particular type illustrates the origin of the units of pressure typically used—millimeters of mercury, or mmHg. A blood pressure of 120 mmHg means a pressure that will support a column of mercury 120 mm high (0.34×). (Photo by D. Morton)

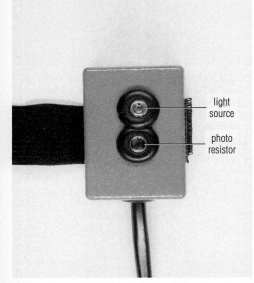

Figure 143b Photoplethysmograph. This sensor for peripheral circulation has two "eyes." One eye emits invisible infrared light, which bounces off red blood cells flowing in finger capillaries. A photoresistor in the other eye collects the reflected light (80×). (Photo by D. Morton)

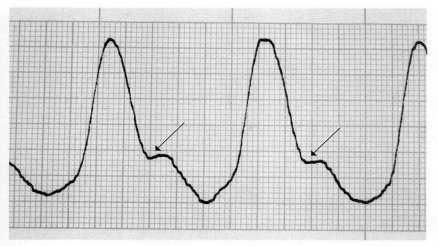

Figure 143c Peripheral blood flow recording. The peaks represent flow during ventricular systole and the troughs, flow during ventricular diastole. The dichrotic notch (arrows) coincides with the slight drop in blood pressure following closure of the aortic semilunar valve. (Photo by D. Morton)

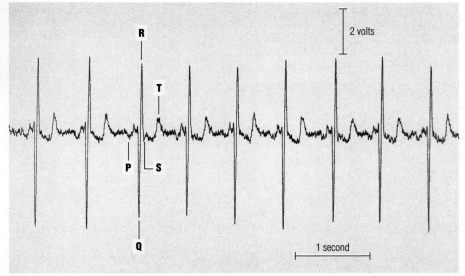

Figure 143d Human electrocardiogram (ECG). The P-wave (inverted in this trace) corresponds to depolarization of the atria. The conduction of the "message to contract" from the atrioventricular node to the ventricular fibers and the depolarization of the ventricles result in the QRS complex of waves. Repolarization of the ventricles produces the T-wave. Repolarization of the atria occurs during the QRS complex and has no particular wave associated with it. (Photo by D. Morton)

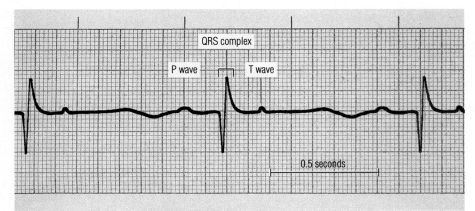

Figure 143e Frog electrocardiogram (ECG). (Photo by D. Morton)

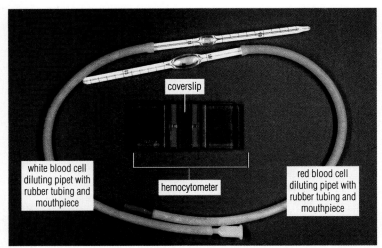

Figure 144a Hemocytometer and blood-diluting pipets. There are several different types of blood-diluting instruments. This type clearly illustrates the calculation of dilution factors. To perform a count, draw blood up to the 0.5 mark. Then draw dilutant up to the mark above the mixing chamber. After gentle mixing, discard the dilutant from the barrel of the pipette. Now place a drop of the diluted blood in one of the frosted triangular areas located above and below the edge of the hemocytometer cover slip. Multiply the actual cell count by a dilution correction factor of $(11 - 1)/0.5 = 20$ for white blood cells and $(101 - 1)/0.5 = 200$ for red blood cells (0.40×). (Photo by D. Morton)

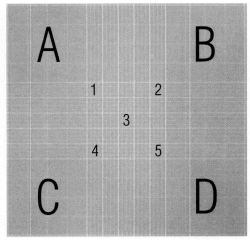

Figure 144b Hemocytometer counting grid. Letters mark the four sets of 16 medium squares used to count white blood cells. The total volume of diluted blood counted is 0.4 cubic millimeters (cu mm). Thus the volume correction factor is 2.5. Numbers mark the five sets of 16 small squares used to count red blood cells. The total volume counted is 0.02 cu mm, producing a volume correction factor of 50. Considering both the dilution and volume correction factors, multiply white blood cell counts by $20 \times 2.5 = 50$ and red blood cell counts by $200 \times 50 = 10,000$ (25×). (Photo by D. Morton)

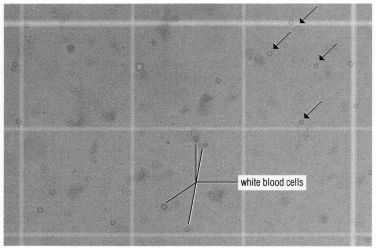

Figure 144c White blood cell count. This is a photograph of the upper right-hand side of large square "B" in Figure 144b. For each medium-sized square, count the cells touching the top and left-hand sides and do not include those touching the other sides. Arrows indicate the four cells counted in the upper right-hand medium square (120×). (Photo by D. Morton)

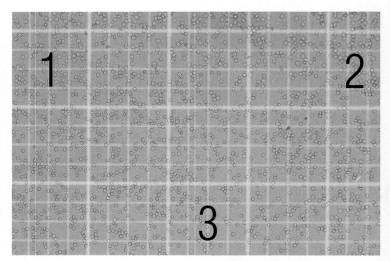

Figure 144d Red blood cells in counting chamber. This is a photograph of part of the central large square in the hemocytometer grid from Figure 144b. Each of the medium-sized squares contains 16 small squares (100×). (Photo by D. Morton)

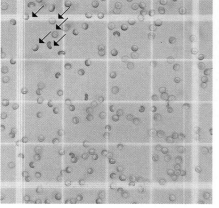

Figure 144e Red blood cell count. This is a photograph of a medium square. The arrows point to the 5 cells counted in the upper left-hand small square. The count in the small square immediately to the right is 8 (250×). (Photo by D. Morton)

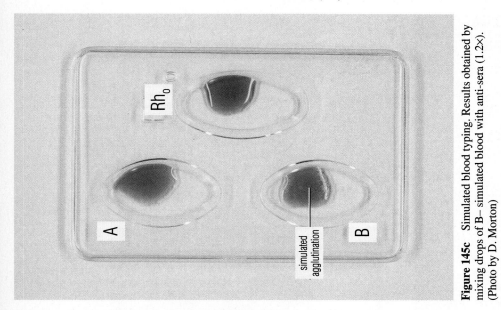

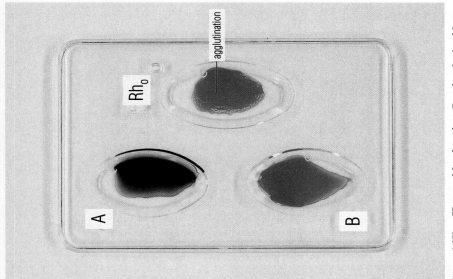

Figure 145a Hematocrit. Centrifugation of blood separates its components. A hematocrit determines the volume of packed red cells. Calculate it by dividing the length of the column of packed red blood cells by the total length of the blood column and multiplying by 100. In this case it is 53% (1.2×). (Photo by D. Morton)

Figure 145b Human blood typing. Results obtained by mixing a drop of test blood with a drop of serum containing antibodies to one of the antigens found on the surface of red blood cells (anti-A, anti-B, or anti-D). If an antigen is present, the red cells clump together (agglutinate). This blood type is O+ (1.2×). (Photo by D. Morton)

Figure 145c Simulated blood typing. Results obtained by mixing drops of B− simulated blood with anti-sera (1.2×). (Photo by D. Morton)

Magnetic Resonance Imaging

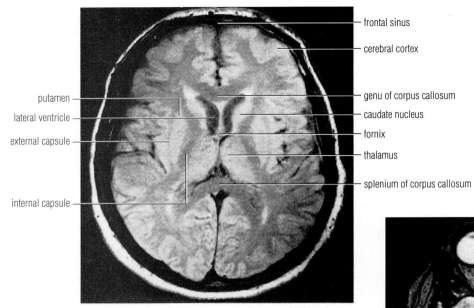

Figure 146a MRI of **transverse section** of **head** at level of **thalamus of brain** (0.40×). (Courtesy Berton Leach)

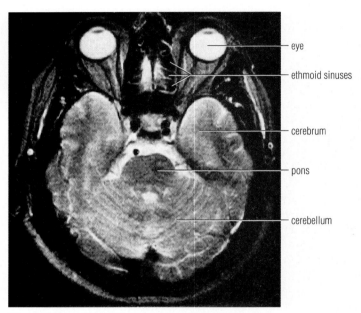

Figure 146b MRI of **transaxial section** of **head** at level of **eyes** and **pons** (0.40×). (Courtesy Berton Leach)

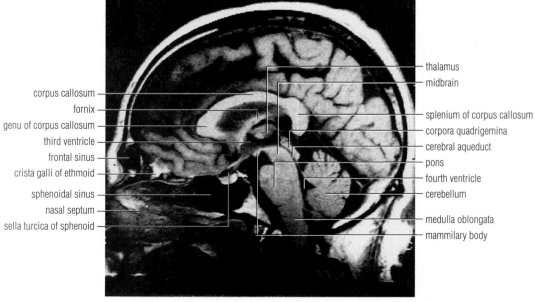

Figure 146c MRI of **midsagittal section** of **head** (0.50×). (Courtesy Berton Leach)

INDEX

abdominal cavity
    cat, 125
    fetal pig, 133
    human, 75
action potential, 138
adipocytes. See fat cells
areolar connective tissue. See loose connective tissue
astrocytes, 25

bacteria, 8
bile canaliculi, 82
blood
    cell counts, 145
    cells, 93
    flow, 143
    platelets, 93
    typing, 146
bone marrow, 94
bone tissue
    canaliculi, 16
    compact, 16
    decalcified, 17
    Haversian canal, 16
    Haversian system, 16
        development, 17
    interstitial lamellae, 16
    lacunae, 16
    lamellae, 16, 17
    osteoblasts, 17–19
    osteoclasts, 17–19
    osteocytes, 16–19
    spicules, 19
    trabecula, 19
bones
    cat, 110
    growth
        endochondral ossification, 18–19
        intramembranous ossification, 19
    human, 34
        hand, 43
        hyoid, 32, 41
        lower appendage, 48–49
        pectoral girdle, 42
        pelvic girdle, 45–47
        rib, 41
        sacrum, 41
        skull
            adult, 33–39
            fetus, 36
        sternum, 41
        upper appendage, 43–44
        vertebrae, 40
Bowman's capsule, 9

brain
    fetal pig, 137
    human
        external anatomy, 63
        microanatomy of cerebellum, 26
        microanatomy of cerebrum, 25
        model, 61
        sagittal section, 62–63
    MRI, 147
    sheep
        coronal (frontal) section, 64
        external anatomy, 64
        sagittal section, 64
brown fat, 14
Brunner's glands, 80

capillaries
    section
        lungs, 85
        thyroid, 71
    whole mount, 92
cartilage
    capsule, 15
    chondrocytes, 15
    extracellular matrix, 15
    growth
        appositional, 15
        interstitial, 15
    isogenous group, 15
    lacuna, 15
cat
    fetus in utero, 131
    internal anatomy, 122–131
    skeletal muscles, 112–121
    skeleton, 110
    skinned, 111
cell
    bacterium, 8
    bone
        osteoblast, 17–19
        osteoclast, 17–19
        osteocyte, 16–19
    cardiac muscle, 22
    cartilage
        chondroblast, 15
        chondrocyte, 15
    chief, 78
    fat, 13
    goblet, 10
    human cheek, 9
    in mitosis, 6–7
    interstitial cells, 101
    macrophage, 8

    megakaryocyte, 94
    Panath, 80
    parietal, 78
    Purkinje, 26
    pyramidal, 25
    skeletal muscle (myofiber), 21
    ultrastructure, 4, 5
chromosomes
    mitosis, 6–7
    ultrastructure, 6
cilia, 10
ciliary body, 67
circulatory system
    gross anatomy
        cat, 123–129
        fetal pig, 134–136
        human heart, 86
        sheep heart, 88
    microanatomy
        blood vessels, 90–92
        heart, 89
cleavage furrow, 7
cochlea, 69
collagen fibers, 12, 15
colon
    gross anatomy
        cat, 125
        fetal pig, 133
        human, 75
    microanatomy, 81
connective tissue
    bone, 16
    brown fat, 14
    cartilage, 15
    dense irregular fibrous, 12
    dense regular
        elastic, 13
        fibrous, 12
    embryonic, 12
    loose, 12
    reticular, 14
    white fat, 13
corpus
    albicans, 105
    luteum, 105
corpuscle
    Meissner's, 29
    Pacinian, 29
cranial nerves, 61

dense irregular fibrous connective tissue, 12
dense regular elastic connective tissue, 13

dense regular fibrous connective tissue, 12
diaphragm
    cat, 124
    fetal pig, 133
    human, 75
digestive system
    gross anatomy
        cat, 122, 123, 125–127
        fetal pig, 132, 133
        human abdominal cavity, 75
        human opened stomach, 76
    microanatomy
        anal canal, 81
        appendix, 81
        colon, 81
        esophagus, 77
        gallbladder, 81
        liver, 82
        pancreas, 82
        salivary glands, 73
        small intestine, 80
        stomach, 77–79
        teeth, 74
        tongue, 69, 72
ductus deferens. See spermatic duct
dura mater, 62

ear
    cochlea, 69
    live, 68
    models, 68
    organ of Corti, 69
elastic cartilage, 15
elastic fenestrated membranes, 13, 90
electrocardiogram (ECG), 144
electroencephalogram (EEG), 138
electromyogram (EMG)
    relationship with force output, 140
    typical, 140
embryonic connective tissue, 12
endocrine system
    adrenal gland, 71
    parathyroid gland, 71
    pineal gland, 71
    pituitary gland, 70
    thyroid gland, 71
endomysium, 21
endoneurium, 27, 28

endosteum, 17
epimysium, 21
epineurium, 27
epiphyseal plate, 19
epithelia
   parenchyma, 11
   pseudostratified columnar, 10
   simple columnar, 10
   simple cuboidal, 9
   simple squamous, 9
   stratified cuboidal, 10
   stratified squamous, 10
      keratinized, 11
   transitional, 11
esophagus
   gross anatomy
      cat, 122
      fetal pig, 133
      human, 62
   microanatomy, 77
eye
   fetus
      sagittal section, 67
   human
      live, 65
      models, 65
   monkey
      ciliary body, 67
      iris, 67
      retina, 67
   sheep, 66

fascia lata, 59
fascicles
   of nerve, 27
   of skeletal muscle, 21
fat cells
   multilocular, 14
   unilocular, 13
fetal pig
   external anatomy, 132
   extraembryonic membranes, 108
   internal anatomy, 133–135, 137
   respiratory system, 83
fibers
   collagen, 12, 15
   elastic, 12–13, 15
   reticular, 14
fibroblasts, 12–13
fibrocartilage, 15
follicles
   hair, 30–31
   ovarian, 105

gallbladder
   cat, 127
   human, 76
ganglion
   spinal, 23, 29
   sympathetic, 29
germinal center, 96
glomerular capsule. *See* Bowman's capsule
glomerulus, 9
glycogen, 82
goblet cells, 10
Gram's stain, 8
gray matter, 23

hair, 30, 31
Hassall's corpuscles, 95
Haversian system, 16–17
heart
   cat, 124
   fetal pig, 134
   human, 86
   microanatomy, 89
   model, 87
   sheep, 88
hematocrit, 146
hemocytometer, 145
hepatic lobules, 82
hepatic portal vein
   cat, 127
   human, 76
hepatocytes, 82
hyaline cartilage, 15

iliotibial tract, 58
integumentary system. *See* skin
intercalated discs, 22
interstitial cells, 101
iris, 67
islets of Langerhans, 82

joint
   ankle, 49
   elbow
      bones, 43
      model, 50
   hip
      bones, 45
      model, 51
   knee
      bones, 48
      model, 52
      microanatomy, 51
   shoulder
      bones, 42
      model, 50
   wrist, 43
juxtaglomerular apparatus, 98

kidney
   gross anatomy
      cat, 129
      fetal pig, 136
      sheep, 97
   microanatomy, 97–98

larynx
   cat, 122–123
   fetal pig, 84
   human, 62
   model, 84
Leydig cells. *See* interstitial cells
liver
   gross anatomy
      cat, 127
      fetal pig, 133
      human, 75–76
   microanatomy, 82
loose connective tissue, 12
lungs
   cat, 122
   fetal pig, 83
   human, 83
lympatic nodules, 96
lymph node
   gross anatomy, 112, 126
   microanatomy, 96
lymphatic system
   bone marrow, 94
   lymph node, 96
   spleen, 95
   thymus, 95
   tonsils, 96
lymphatic vessels
   lacteal, 80
   valve, 92

macrophage, 8
macula densa, 98
magnetic resonance imaging (MRI), 147
mammary gland, 109
megakaryocyte, 94
mesentery
   cat, 126
   human, 75
microscope
   compound, 1
   parts, 3
   dissecting, 2

mitosis
   HeLa cells, 7
   ultrastructure, 6
   whitefish blastula, 6–7
models
   brain, 61
   ear, 68
   elbow joint, 50
   eye
      disassembled, 65
      intact, 65
   female reproductive system, 104
   heart, 87
   hip joint, 51
   knee joint, 52
   larynx, 84
   male reproductive system, 100
   shoulder joint, 50
motor end plates, 28
motor unit recruitment, 141
mucous connective tissue. *See* embryonic connective tissue
muscle contraction
   electromyogram (EMG), 140
   fatigue, 142
   length, 141
   summation, 142
   tetanus, 142
   treppe, 141
   twitch, 142
   twitch fusion, 142
muscle tissue
   cardiac, 22
   skeletal, 20–21
   smooth, 22
muscular system. *See* skeletal muscles
myelin sheaths, 27, 28
myofibrils, 20, 21
myofilaments, 21

nephron, 9, 97, 98
nerve
   gross anatomy, 54–55, 57, 59–60
   microanatomy, 27–28
nervous system
   gross anatomy
      fetal pig, 137
      human, 62–63
      sheep brain, 64

microanatomy
   cerebellum, 26
   cerebrum, 25
   nerve, 27
   spinal cord, 23
   spinal ganglion, 29
   sympathetic ganglion, 29
   model of human brain, 61
nervous tissues
   accessory cells
      astrocytes, 25
      neroglia, 24
      Schwann cells, 28
   myelin sheaths, 27–28
      nodes of Ranvier, 28
      Schmidt-Lanterman clefts, 28
   neuron
      somatic motor, 24, 28
      structure, 24
nipple, 109

olfactory epithelium, 69
omentum
   greater, 125
   lesser, 126
optic disk, 67
oral cavity
   cat, 122
   fetal pig, 132
organ of Corti, 69
osteon. See Haversian system
ovary
   gross anatomy
      cat, 130
      fetal pig, 137
   microanatomy, 104–105
oviduct. See uterine tube

pancreas
   gross anatomy
      cat, 127
      human, 76
   microanatomy, 82
parenchyma, 11
patella reflex, 138
perichondrium, 15
perimysium, 21
perineurium, 27
periosteum, 17
pharynx
   cat, 122
   fetal pig, 132
photoplethysmograph, 143
pithing a frog, 139
placenta, 108

portal lobules, 82
portal triads, 82
pseudostratified columnar epithelium, 10
Purkinje cells, 26
Purkinje fibers, 89
pyloric sphincter, 79
pyramidal cells, 25

refractory period
   absolute, 138
   relative, 138
renal corpuscle, 9
reproductive system
   gross anatomy
      cat, 130
      fetal pig, 136
   microanatomy
      cervix, 107
      epididymis, 102
      ovary, 104
      penis, 103
      prostate, 103
      seminal vesicle, 102
      sperm, 102
      spermatic cord, 103
      spermatic duct, 103
      testis, 101
      uterine tube, 106
      uterus, 106–107
      vagina, 107
   model of female, 104
   model of male, 100
respiratory system
   gross anatomy
      cat, 123
      fetal pig, 83–84, 132–133
      human, 62, 83
   microanatomy
      lung, 85
      trachea, 85
   model of human larynx, 84
rete testis, 101
reticular connective tissue, 14
reticular fibers, 14
retina, 67

salivary glands
   parotid, 73
   sublingual, 73
   submandibular, 73
sebaceous glands, 31
seminiferous epithelium, 101
serous demilune, 73
simple columnar epithelium, 10
simple cuboidal epithelium, 9

simple squamous epithelium, 9
skeletal muscle tissue
   A-bands, 20–21
   filaments, 21
   H-zone, 21
   I-bands, 20–21
   M-line, 21
   myofibrils, 20–21
   sarcomere, 20–21
   striations, 20
skeletal muscles
   cat
      abdominal, 118
      deep muscles of upper back, 117
      deeper muscles of thigh, 121
      deeper muscles of chest, 114
      deeper muscles of forearm, 113
      deeper muscles of lateral upper arm, 116
      deeper muscles of upper arm, 114
      deeper muscles of upper back, 116
      intercostals, 114
      lower back, 117
      muscles of lower leg, 120
      neck, 112
      superficial muscles of leg, 119
      superficial muscles of upper back, 115
      superficial muscles of chest, 113
      superficial muscles of dorsal neck, 115
      superficial muscles of lateral arm, 115
      superficial muscles of leg, 120
      superficial muscles of medial arm, 113
      supraspinatus, 116
   fasicles, 21
   human
      abductor pollicis, 56
      adductor longus, 59
      adductor magnus, 58
      anconeus, 56
      biceps brachii, 55–56
      biceps femoris, 58–60
      brachioradialis, 55–56
      buccinator, 53

deltoid, 55
depressor anguli oris, 53
depressor labii inferioris, 53
digastric, anterior belly, 54
extensor carpi radialis brevis, 56
extensor carpi radialis longus, 56
extensor carpi ulnaris, 56
extensor digiti minimi, 56
extensor digitorum, 56
extensor digitorum brevis, 60
extensor digitorum longus, 60
extensor hallucis longus, 60
extensor pollicis brevis, 56
external oblique, 55–58
flexor carpi radialis, 56
flexor carpi ulnaris, 56
frontalis, 53
gastrocnemius, 59–60
geniohyoid, 54
gluteus maximus, 58–59
gluteus medius, 58
gracilis, 58–60
hyoglossus, 54
infraspinatus, 55
internal oblique, 57
lateral head of triceps brachii, 55
latissimus dorsi, 55
levator labii superioris, 53
levator scapulae, 53–54
masseter, 53
mentalis, 53
mylohyoid, 54
nasalis, 53
occipitalis, 53
omohyoid, inferior belly, 54
omohyoid, superior belly, 54
orbicularis oculi, 53
orbicularis oris, 53
pectoralis major, 55
peronus longus, 60
pronator teres, 56
pyramidalis, 57
rectus abdominis, 57
rectus femoris, 59

skeletal muscles
  human *(continued)*
    rhomboideus major, 55
    risorius, 53
    sartorius, 59–60
    scalenus anterior, 54
    scalenus medius, 54
    semimembranosus, 58–60
    semitendinosus, 58–60
    serratus anterior, 55
    splenius capitis, 53
    sternocleidomastoid, 53–54
    sternohyoid, 54
    sternothyroid, 54
    temporalis, 53
    tensor fascia latae, 59
    teres major, 55
    teres minor, 55
    thyrohyoid, 54
    tibialis anterior, 60
    transversus abdominis, 57
    trapezius, 54–55
    triceps brachii, 55
    vastus lateralis, 59
    vastus medialis, 59
  microanatomy, 21
  sheep eye, 66
skeletal system
  cat, 110
  human, 32–48
skeleton
  cat, 110
  human
    appendicular, 42–48
    axial, 33–41
skin
  epidermal strata, 30
  layers, 30
small intestine
  gross anatomy
    cat, 125
    fetal pig, 133
    human, 75
  microanatomy, 22, 80
spermatic duct, 103
spermatogenesis, 101
sphygmomanometer, 143
spinal cord
  gross anatomy
    fetal pig, 137
    human, 62
  microanatomy, 23
  smear, 24
spindle, 6

spiral organ. *See* organ of Corti
stomach
  cat, 126
  gross anatomy
    fetal pig, 133
    human, opened, 76
  microanatomy, 77–79
    cat, opened, 126
stratified cuboidal epithelium, 10
stratified squamous epithelium, 10
  keratinized, 11
summation, 142
sweat glands, 31
  ducts, 10

taste bud, 69
tetanus, 142
thoracic cavity
  cat, 122
  fetal pig, 133
  human, 83
thrombocytes, 93
tongue
  circumvallate papillae, 72
  filiform papillae, 72
  foliate papillae, 69
  fungiform papillae, 72
  live, 72
tooth
  development, 74
  microanatomy, 74
trachea, 15
transitional epithelium, 11
treppe, 141
twitch, 142
twitch fusion, 142

ultrastructure
  bacteria, 8
  cell organelles, 4–5
  chromosomes, 6
  mitosis, 6
  skeletal muscle
    fiber, 21
    fibril, 21
umbilical cord, 108
urinary system
  gross anatomy
    cat, 129
    fetal pig, 136
    sheep kidney, 97
  microanatomy
    kidney, 97–98
    ureter, 99

    urethra, 99
    urinary bladder, 99
uterine tube, 106
uterus
  menstrual phase, 107
  progravid phase, 107
  secretory phase, 106

vas deferens. *See* spermatic duct
Volkman's canal, 16

white fat, 13
white matter, 23